Virology of Flowering Plants

TERTIARY LEVEL BIOLOGY

A series covering selected areas of biology at advanced undergraduate level. While designed specifically for course options at this level within Universities and Polytechnics, the series will be of great value to specialists and research workers in other fields who require a knowledge of the essentials of a subject.

TERTIARY LEVEL BIOLOGY

Virology of Flowering Plants

W. A. STEVENS, B.Sc., Ph.D.

Lecturer in Botany
Royal Holloway College
University of London

Springer Science+Business Media, LLC

Blackie & Son Limited,
Bishopbriggs, Glasgow G64 2NZ

Furnival House, 14–18 High Holborn, London WC1V 6BX

Distributed in the USA by
Chapman and Hall
in association with Methuen, Inc.
733 Third Avenue, New York, N.Y. 10017

© 1983 Springer Science+Business Media New York
Originally published by Blackie & Son Ltd in 1983.
Softcover reprint of the hardcover 1st edition 1983
First published 1983

British Library Cataloguing in Publication Data
Stevens, W. A.
Virology of flowering plants.—(Tertiary level
biology)
1. Virus diseases of plants
I. Title
582.13′04234   SB736

ISBN 978-1-4757-1253-7   ISBN 978-1-4757-1251-3 (eBook)
DOI 10.1007/978-1-4757-1251-3

For the USA, International Standard Book Numbers are
978-0-412-00061-4
978-0-412-00071-3 (pbk)

Filmset by Advanced Filmsetters (Glasgow) Ltd

Preface

Knowledge of plant viruses has increased markedly in the last few years, mainly as a result of intensified research by both plant pathologists and molecular biologists. A stimulus to the interest in plant viruses has been the finding that some share closely many of the structural characteristics and features of replication associated with viruses of animals and bacteria. The broadening of our knowledge of viruses has meant that increasing reference is made in biological sciences to plant viruses, and there appears now to be a need for a general text which is suitable for undergraduates, and which will also serve as an introduction for postgraduates where an understanding of plant virology is perhaps peripheral to their research or teaching interests. It is to meet this demand that this book has been written.

It is hoped that chapters on symptoms, transmission and control of plant virus diseases will prove particularly interesting to student plant pathologists and agriculturalists, while the accounts of structure and virus replication will be of importance to micro- and molecular biologists. An outline of the numerous techniques employed in studies of viruses is given in the final chapter.

The bibliography includes books and reviews of general interest in plant virology. A further list, sub-divided by chapters, directs the student towards the relevant specialist literature.

I would like to record my sincere gratitude to my wife and family for their support and encouragement during the preparation of this book. I dedicate this work to them. I would also like to thank Mrs Inge Judd who patiently typed and retyped the manuscript. I gratefully acknowledge the help and skill of Mr David Ward who took most of the photographs. Thanks are due to Miss Carole Spurdon who was responsible for the preparation of materials for, and operation of, the electron microscope, and to Mrs Diana Hughes for typing the legends and labels for the diagrams. To all my other colleagues and friends who have given me encouragement, advice and help, I am sincerely grateful.

Where line illustrations are based on previous published work, and where photographs are supplied by colleagues, acknowledgment has been included in the legend.

W.A.S.

Contents

CHAPTER ONE

INTRODUCTION

1.1 Brief history of plant virology

In the year AD 752 the Japanese empress KoKen wrote a poem about the
beautiful yellow leaf of *Eupatorium chinensis*. This may well be the first
description in literature of a plant virus disease. Nettlehead, a condition in
hops, now known to be due to virus infection, was first described by Scot in
1574, but the earliest known pictorial records of plant viruses are the
'broken' or Rembrandt tulips often depicted in paintings from the 17th
century Dutch school of art.

These three examples show that unwitting records of the effects of
viruses on plants have been known for many years. However, the science of
plant virology may be thought to have originated in 1882 when Adolf
Mayer, Professor of Chemical Technology and later Professor of Botany
at Heidelberg University, described a 'mosaic' disease of tobacco. Mayer
demonstrated that the discoloration or mosaic condition could be transferred to healthy tobacco by rubbing leaves with sap from the mosaic
plants. Ivanowsky, a Russian botanist, noticed Mayer's work and
recognized the same disease condition in Ukrainian and Bessarabian
tobacco crops. He went on to show in 1892 that the principle producing
disease retained activity even after passing through a 'bacteria-proof' filter.
Ivanowsky, influenced by the knowledge and thinking of his day, identified
the pathogen as a toxin-producing entity. A Dutch microbiologist, Jan
Beijerinck, repeated and expanded the work of Ivanovsky by showing that
the mosaic agent multiplied in plant tissue and could not therefore be a
toxin. In 1898 Beijerinck called the agent a 'contagium vivum fluidum'.
This was a revolutionary idea at the time, since substances were considered
to be either 'corpuscular' (such as bacteria and blood cells), or 'dissolved',

1

like salts and other small molecules in solution. The idea that the disease agent was fluid, and therefore dissolved, seemed extraordinary when at the same time it was said to be living and capable of reproducing.

From about 1890 the term 'virus' was used to denote a wide range of infectious agents including bacteria and unidentified microbes. When some disease agents were found to pass through filters designed to stop bacteria, they became known as 'filterable viruses'. This term was also used for a number of agents of animal diseases including foot-and-mouth, rabies and fowl pox. As late as 1928, Boycott wrote an article entitled 'The transition from live to dead: the nature of filterable viruses'. Workers trying to estimate the size of the tobacco mosaic agent described it as a virus in 1921. Gradually 'virus' was used to denote a distinct group of submicroscopic disease agents. The invisibility of these agents was very baffling to early workers, and Maurice Mulvania, a biochemist, suggested in 1926 that the mosaic virus of tobacco might be a 'protein of a very simple kind having characteristics of an enzyme'. In the same year James Sumner, an organic chemist, had crystallized the enzyme urease from jack bean (*Canavalia ensiformis*). Since the mosaic agent was thought to be enzyme-like, the possibility of purifying and crystallizing tobacco mosaic virus (TMV) attracted the attention of biochemists. Eventually in 1935 the American, Wendell Meredith Stanley, announced the isolation of a 'crystalline protein possessing the properties of tobacco mosaic virus'. Work led by Bawden and Pirie in Britain (1936) showed that TMV formed liquid crystals rather than true crystals, and also that pure preparation contained nucleic acid. By 1939 several plant viruses had been shown to consist of nucleoprotein. By physical methods some viruses like TMV were shown to be rod-shaped and others spherical, and these morphological features were confirmed and expanded upon when the electron microscope was developed in the 1940s.

Because only 5% of the TMV particle consisted of ribonucleic acid (RNA), this was considered to be a minor virus component. In 1956, however, Gierer and Schramm from Tübingen and Fraenkel-Conrat in California showed simultaneously and independently that the RNA of TMV could alone cause infection. This result complemented the earlier findings of workers using viruses of bacteria (bacteriophages), who showed that only the bacteriophage nucleic acid enters the host bacterium. These studies of TMV proved to be of enormous importance and significance, not only to the study of both plant and animal viruses, but also to fundamental research in molecular biology and genetics.

Chemists and biochemists were not the only scientists preoccupied with research into viruses. Plant pathologists were busy describing symptoms and identifying new virus diseases or suspected virus diseases. Studies of virus disease outbreaks (epidemiology) drew attention to the mechanisms of virus spread. Transmission of rice dwarf virus by leaf-hoppers was established in 1893 and confirmed experimentally in 1937. Other vectors such as aphids and mites were implicated in virus transmission about this time. In 1958 it was confirmed that virus transmission through the soil came about in part by nematode worms (eelworms). Later (1960) fungi were also shown to transmit viruses through the soil from plant to plant.

1.2 What is a virus?

As more information has accumulated concerning the chemical and physical characteristics as well as replicative features of viruses, so changes have taken place in the definition of a virus. Bawden (1956) defined a virus as an obligate parasitic pathogen with dimensions of less than 200 nm. Although possibly adequate in its day, such a definition does not exclude naked nucleic acid pathogens—the viroids, or some mycoplasmas. The more information is gathered about viruses and other submicroscopic organisms, the more difficult it appears to give a precise statement of the term 'virus'. Discussions on the accuracy and formulation of definitions of viruses are given by Luria et al. (1978) and by Matthews (1970, 1981).

Table 1.1 Sizes of some plant viruses.

	Virus	Dimensions (nm)
Isometric (spherical)		
	Satellite	17
	Tobacco necrosis	26
	Cucumber mosaic	30
	Cauliflower mosaic	50
	Tomato spotted wilt	80
Rod-shaped		
	Tobacco mosaic	300×18
	Potato virus X	520×13
	Potato virus Y	740×12
	Beet yellows	1200×10
	Citrus tristeza	2000×12
Bacilliform		
	Lettuce necrotic yellows	230×70

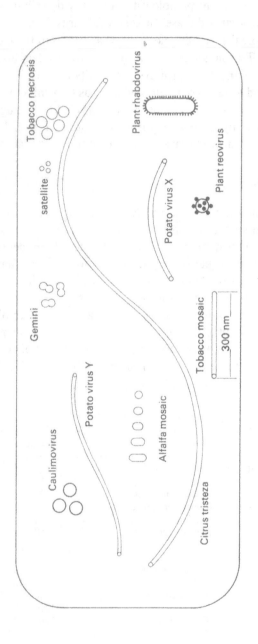

Figure 1.1   Sizes and shapes of plant viruses (outer frame represents a typical bacterial cell of *Escherichia coli* for comparison).

For our purposes a definition utilizing the essential characteristics of viruses will suffice. These essential features are as follows.

(1) Viruses contain one or more pieces of a single type of nucleic acid (RNA or DNA).
(2) The nucleic acid is coated with one or more layers of protein molecules.
(3) Viruses rely on living host cells for most of the enzymes necessary for their replication.

Viruses are therefore submicroscopic particles made of one or more pieces of a single species of nucleic acid, RNA or DNA, surrounded by proteins. These particles replicate alone or in the presence of similar structures, but only in living cells, using at least some of the host cell enzymes. This definition draws attention to the fact that some viruses replicate only in the presence of other viruses. Viroids—naked nucleic acid pathogens and mycoplasms which contain both RNA and DNA—are excluded by this definition.

The reference to size is not very precise but shows that viruses are too small to be seen in the light microscope. Table 1.1 lists some viruses and gives an indication of their size. The dimensions of viruses are usually determined by electron microscopy and measured in nanometres (nm) (see Fig. 1.1). One nanometre is a millionth part of a millimetre. An angstrom (Å) is one-tenth of a nm. To gain some idea of the small size of viruses let us imagine that a 25-nm diameter virus particle is magnified in the electron microscope by 40000—it will then appear 1.0 mm in diameter. If it were possible to magnify, say, the 5-mm diameter head of a nail by this same amount it would appear 200 m (218 yards) in diameter!

1.3 Are viruses organisms?

An organism may be defined as an entity composed of interdependent parts that show characteristic features of life, and which can reproduce either on its own or by interaction with a similar entity to produce a continuous lineage of individuals with the potential for evolutionary change.

The characteristic features of life that distinguish living from non-living matter are the ability to (1) assimilate (metabolize) with the release of energy, (2) excrete waste products of metabolism, (3) grow, (4) reproduce, and also (5) to exhibit some form of irritability or response to the environment. Viruses do not show these characteristics of life, although in other respects they have features in common with organisms as defined above. Viruses cannot replicate on their own—they are obligate parasites, requiring the enzymes and ribosomes of other organisms. Viruses, like viroids, appear therefore to be unique structures.

Table 1.2 Some major groups of organisms invaded by viruses.

Protozoa	Aves (birds)	Algae
Insecta	Mammalia	Fungi
Pisces (fish)	mycoplasmas	Pteridophyta (e.g. ferns)
Amphibia	spiroplasmas	Gymnospermae (e.g. conifers)
Reptilia	Bacteria	Angiospermae (flowering plants)

1.4 Virus hosts

Viruses enter and multiply in a wide range of hosts—Table 1.2 shows the diversity of organisms invaded by viruses. In this book, only viruses of flowering plants will be considered. Flowering plants include plants of economic importance, such as ornamentals grown for pleasure and profit as well as a very wide range of crop plants, from cereals and root crops to fruit trees and bushes. Viruses have been recorded from numerous families of flowering plants, but naturally those from plants of economic importance have been studied most and are therefore best documented.

1.5 Terminology

Most of the terms used in virology will be explained as they are used. It is useful, however, to introduce some general terms at this point (Fig. 1.2).

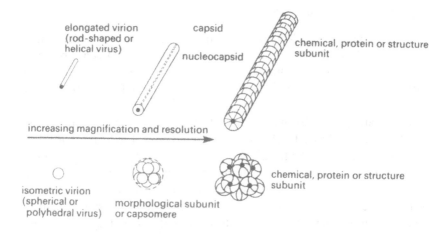

Figure 1.2 Diagram to illustrate some terms used in plant virology.

The external shape of the virus as it appears in the electron microscope may be (*a*) elongated with *helical symmetry* or (*b*) apparently spherical or *isometric*. Viruses with helical symmetry are also known as anisometric, tubular, filamentous, or rod-shaped; isometric viruses are also known as polyhedral or icosahedral.

Viruses with combined symmetry or with geometry other than that based on helical or isometric symmetry may be called *complex*. Some have parallel sides, with one or both ends rounded, and are known as *bacilliform*.

The intact virus particle is known as the *virion*. The nucleic acid of the virion plus the protein in contact with it comprises the *nucleocapsid*.

The protein coat, without the nucleic acid, is the *capsid* and is composed of structural units. In isometric viruses structural units may be grouped together to make morphological units or *capsomeres*, which are large enough to be resolved in the electron microscope. In some viruses the nucleocapsid may be enclosed in a layer or layers of protein called the *envelope*.

1.6 Virus names and groups

Each virus is named according to the major host with which it is associated and also the symptoms produced in that host. However, a plant virus may invade a number of different plant species, producing different symptoms in each, and the identity of the major host may not always be clear, so that a virus may have many names. Cucumber mosaic virus (CMV) illustrates this point well—Table 1.3 shows the range of symptoms produced in five different hosts.

The fern leaf condition induced by this virus in tomato, for example, has led to descriptions of 'tomato fern leaf disease' and 'tomato fern leaf virus' appearing in the literature. Later it was realized that the causative agent was CMV. In fact, Martyn (1968, 1971) has listed over 80 diseases from 70 plant species induced by CMV. A further complication is that the

Table 1.3 Symptoms produced by cucumber mosaic virus in various hosts.

Host plant	Symptoms
Cucumber (*Cucumis sativus*)	Green or yellow mosaic
Tobacco (*Nicotiana tabacum*)	Systemic mosaic or ringspots
Tomato (*Lycopersicon esculentum*)	Narrow leaf lamina (fern leaf)
French bean (*Phaseolus vulgaris*)	Small pin-point brown (necrotic) lesions
Cowpea (*Vigna sinensis*)	Larger brown spots (local lesions)

Table 1.4 Symptoms produced by type strain tobacco mosaic virus in tobacco varieties.

Nicotiana tabacum var. Samsun White burley Xanthi Turkish	Vein clearing followed by mosaic, leaf distortion and blistering. Symptoms suppressed by limited nitrogen supply.
N. tabacum var. Samsun NN *N. tabacum* var. Xanthi—nc	Necrotic spots (local lesions) at temperatures below 28°C. Systemic at higher temperatures.
N. tabacum var. Java	Systemic infection, no vein clearing.

symptoms produced by any plant virus combination depend on the age and physiology of the host plant, climate and also on environmental conditions. Furthermore, symptoms vary with slight variation in the virus (virus strain) and in host plants (plant variety). Table 1.4 shows how different symptoms are produced by TMV in different varieties of tobacco.

It seems obvious that more fundamental information is required for the adequate and precise naming of each virus. On the other hand, host plant

Figure 1.3 Tobacco mosaic virus—a rigid rod or elongated virus.

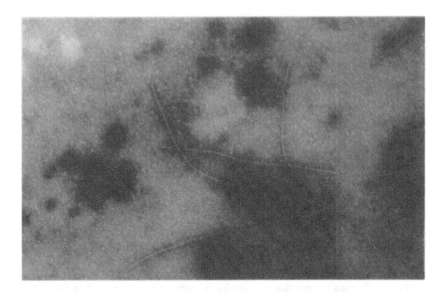

Figure 1.4 Potato virus X—a flexuous rod or elongated virus.

identity and symptoms produced are the most easily assessable pieces of information available for naming purposes. With this in mind it is now the practice to retain the common or vernacular name, but to quote, where possible, a code giving some properties of the virus. This code or *cryptogram* can be used to describe viruses infecting any group of organisms, and is based on four pairs of symbols. For example, the cryptogram for TMV is R/1: 2/5: E/E: S/O, the symbols having the following meaning.

The first pair indicates the *type* of nucleic acid and *strandedness* of the nucleic acid. The symbols for type of nucleic acid are R = RNA; D = DNA. Those for strandedness are 1 = single-stranded; 2 = double-stranded.

The second pair gives information concerning the *molecular weight* of nucleic acid (genome) (in millions) and the *percentage* of nucleic acid in infective particles. The nucleic acid (genome) of some viruses is in several pieces. Where these occur together in one type of particle, the symbol Σ indicates the total molecular weight of the genome in the particle (e.g. clover wound tumour virus, R/2: $\Sigma16/22$:S/S:S, I/Au), but when the pieces of the genome occur in different particles, the composition of each

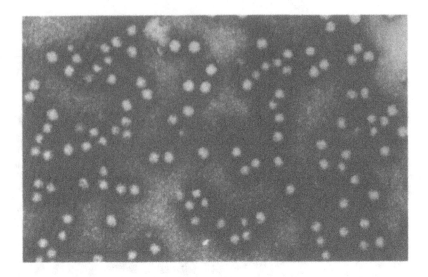

Figure 1.5 Tobacco necrosis virus—a spherical or isometric virus. Particle diameter 26 nm (photograph kindly supplied by Roger Turner, Rothamsted Experimental Station).

particle type is listed separately. The cryptogram for cucumber mosaic virus is

$$R/1:\frac{1.3}{18}+\frac{1.1}{18}+\frac{0.8+0.3}{18}:S/S:C.V/Ap.$$

This shows that four pieces of RNA are encapsidated in three particles.

The third pair of symbols indicates the *outline* of the particle and of the nucleocapsid. Symbols for both these properties are:

S = essentially spherical
E = elongated with parallel sides, ends not rounded
U = elongated with parallel sides, end(s) rounded
X = complex or none of the above.

The last pair of symbols shows the kind of *host* infected and the nature of the *vector* transmitting the virus. The symbols for kinds of host are:

A = alga   M = mycoplasma
B = bacterium   P = pteridophyte
F = fungus   S = seed plant
I = invertebrate   V = vertebrate

The symbols for kinds of vector are:

Ac = mite and tick (Acarina)
Al = whitefly (Aleyrodidae)
Ap = aphid (Aphididae)
Au = leaf-, plant-, or tree-
   hopper (Auchenorrhyncha)
Cc = mealy bug (Coccidae)
Cl = beetle (Coleoptera)
Di = fly and mosquito (Diptera)

Fu = fungus
Gy = Gymnocerata
Ne = nematode (Nematoda)
Ps = psylla (Psyllidae)
Si = flea (Siphonaptera)
Th = thrip (Thysanoptera)
Ve = vectors none of above
0 = spreads without a vector.

Some symbols may be used for all pairs of characters. These are * which indicates that the property of the virus is not known, and ( ) showing that the enclosed information is doubtful or unconfirmed. If the cryptogram is enclosed in square brackets the information applies to a virus group.

The use of cryptograms, together with comparison of symptom production and means of transmission, has enabled plant viruses to be placed into groups of what seem to be related viruses (Table 1.5). Viruses placed in a group share characters as depicted in the cryptogram, but differ for example in the amino acid composition of their coat proteins; they also differ in host range and symptom production, as well as often being transmitted by different vector species. In some cases, where these differences are only slight, viruses are classed as strains of one another.

Table 1.5 Possible groupings of plant viruses.

Virus group	Cryptogram	Example
Elongated viruses		
1. Tobravirus	R/1:2.3/5:E/E:S/Ne	Tobacco rattle virus
		Pea early browning
2. Tobamovirus	R/1:2/5:E/E:S/*	Tobacco mosaic virus
3. Potexvirus	R/1:2.1/6:E/E:S/(Fu)	Potato virus X
4. Carlavirus	R/1:*/6:E/E:S/Ap	Carnation latent virus
5. Potyvirus	R/1:3.5/5:E/E:S/Ap	Potato virus Y
		Tobacco etch
		Henbane mosaic
6. Closterovirus	R/1:4.3/5:E/E:S/Ap	Beet yellows virus
7. Carlavirus	R/1:*/6:E/E:S/Ap	Potato virus S
8. Hordeivirus	R/1:0.9–1.4/4:E/E:S/	Barley stripe mosaic virus
Isometric viruses		
9. Cucumovirus	R/1:1/18:S/S:S/Ap	Cucumber mosaic virus
10. Tymovirus	R/1:1.9/37:S/S:S/Cl	Turnip yellow mosaic virus
11. Comovirus	R/1:1.5/24+	
	2.1/31:S/S:S/Cl	Cowpea mosaic virus

Table 1.5—*continued*

Virus group	Cryptogram	Example
12. Nepovirus	R/1:2.2/40:S/S:S/Ne	Tobacco ringspot virus
		Grapevine fan leaf
		Arabis mosaic virus
		Tomato ringspot
13. Bromovirus	R/1:1/22:S/S:S/(*)	Brome mosaic virus
		Cowpea chlorotic mottle virus
14. Tombusvirus	R/1:1.5/17:S/S:S/*	Tomato bushy stunt virus
15. Caulimovirus	D/2:4.5/16:S/S:S/Ap	Cauliflower mosaic virus
		Dahlia mosaic virus
16. Geminivirus	D/1:*/*:S/S:S/Au	Maize streak virus
17. Ilarvirus	R/*:*/16:S/S:S/O	Prunus necrotic ringspot virus
18. Luteovirus	R/1:2/*:S/S:S/Ap	Barley yellow dwarf virus
19. Maize chlorotic dwarf virus group	R/1:3.2/(36):S/S:S/Au	—
20. Plant reoviruses (A) Phytoreoviruses	R/2:Σ18/22:2/2:SI/Au	Wound tumour virus
(B) Fijiviruses	R/2:Σ15/*:S/S:SI/Au	Sugar cane fiji disease
21. Southern bean mosaic virus	R/1:1.4/21:S/S:S/Cl	
22. Tobacco necrosis virus	R/1:1.5/19:S/S:S/Fu	
Bacilliform viruses		
23. Plant rhabdoviruses (c.f. rabies and vesicular stomatitis)	R/1:4/2:U/E:S,I,V/ Ap.Au.Di.O	Lettuce necrotic yellows Potato yellow dwarf
24. Alfalfa mosaic virus	R/1:1.1/16+0.8/16+ 0.7/16:U/U:S/Ap	—
Ungrouped viruses		
Cacao swollen shoot virus	*/*:*/*:U/U:S/Cc	—
Tomato spotted wilt virus	R/*:*/*:S/S:S/Th	—

1.7 Importance of virus diseases of flowering plants

The effects of viruses on their host plant may often go virtually unnoticed, although there is some loss of plant vigour resulting in a reduction of quality and yield. In some cases, the effects are very obvious, particularly where virus epidemics occur, as in the case of tristeza infections of citrus in Brazil. Six million citrus trees were killed by outbreaks of this aphid-transmitted virus in 1946. Control of the disease involved destruction and burning of orange groves and replanting with new plants grafted on to virus-tolerant root stocks. Since this initial outbreak, epidemics have also been reported in California, Florida, and Spain, and the virus was reported to have spread to Israel in 1970.

In the eastern region of Ghana, the total production of cocoa fell from 118 000 tonnes in 1936 to 39 000 tonnes in 1955, due mainly to the effects of

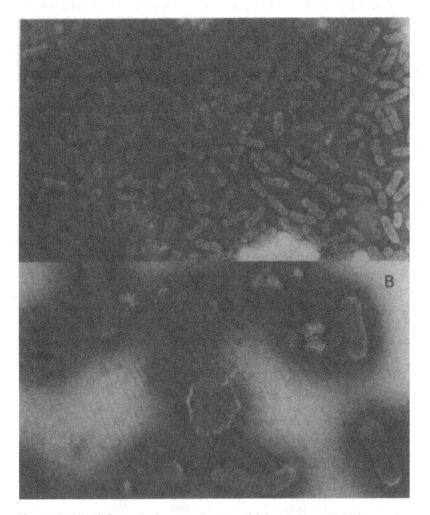

Figure 1.6 (*A*) Alfalfa mosaic virus—maximum particle length 56 nm. (*B*) Maize mosaic—a plant rhabdovirus. Maximum particle length 260 nm. (Photographs kindly supplied by Roger Turner, Rothamsted Experimental Station).

cacao swollen shoot virus (CSSV). Production has remained at this level in spite of replanting (Legg, 1979). The prevalence of this virus disease, which causes drastic loss of production and capital investment in a crop taking four or more years to reach maturity, makes CSSV of particular economic importance in West Africa.

Another crop of considerable economic value in Africa is cassava. About 30 million tonnes of cassava, representing 35% of world production, is grown in Africa. Losses of 80% of yield can be attributed to infection by cassava mosaic virus. This is true particularly of West and East Africa where practically all plants are infected. Losses of 30% throughout Africa are quite common. It seems likely that similar losses may occur in other Third World countries, representing a considerable financial as well as commodity loss.

In the UK it is estimated that 60–70% of cereals are infected with barley yellow dwarf virus (BYDV), resulting in a 5–10% loss of yield. This represents an annual financial loss of £60–120m at 1978 prices (ADAS, 1979).

Another disease of considerable importance in the UK is virus yellows of sugar beet, caused by beet yellows and wild beet yellowing viruses. The financial loss due to virus yellows has been analysed by Heathcote (1978), who shows that a 26% loss of crop yield may result in an even greater percentage financial loss, because reduced plant yield gives a substantial reduction in plant sugar content.

Some virus diseases present international problems. Peppers, for example, are grown commercially in the USA, India, Israel, South America and Japan. There are almost 40 names for different pepper diseases. In the USA, where pepper growing is no longer a minor activity, some areas in the south-west are forced occasionally to abandon pepper crops without

Table 1.6 Effect of four viruses on the yield of peppers (*Capsicum annum*) var. Jalapeno M, from two locations in the USA.

Virus	Percentage loss	
	California	Texas
Tobacco etch	9.1	35.2
Potato virus Y	50.4	58.7
Pepper mottle	43.5	59.1
Tobacco mosaic	42.3	66.7

harvest because of severe virus disease. Table 1.6 shows the degree of loss recorded for different viruses in pepper (Villalon, 1981).

A disease of importance in the USA, particularly in maize-growing areas, is corn lethal necrosis disease (CLND), which results from a combination of maize chlorotic mottle (MCMV) and maize dwarf mosaic virus (MDMV), or MCMV and wheat streak mosaic virus. In Kansas, yield losses of 50–90 % can be attributed to CLND (Uyemoto *et al.*, 1981). Viruses are also responsible for significant losses in yield of glasshouse-grown tomatoes and ornamentals such as chrysanthemums and carnations.

Numerous examples of the importance of viruses in crop production can be found in the literature, but sufficient has been said to emphasize that studies of plant viruses are important in preventing crop losses and generally raising productivity in agriculture and horticulture. Another important aspect of plant virology is its continuing contribution to molecular biology and our understanding of the role of nucleic acid and protein in living organisms.

CHAPTER TWO

SYMPTOMS OF PLANT VIRUS INFECTION

2.1 General features

As several hundred flowering-plant viruses have been described, it is not surprising that there are a very large number of symptoms attributed to these pathogens. Viruses may produce visible or otherwise detectable abnormalities in plants which are recognized as *symptoms of disease*. When no sign of infection can be detected, infection is said to be *latent*. The factors controlling the production of symptoms and their nature are numerous, and include

(1) type and strain of virus
(2) type and variety of host plant
(3) age and stage of development of host
(4) physiology of host
(5) duration of infection
(6) presence of other viruses and pathogens
(7) environmental and climatic conditions.

A given virus may multiply in a number of different plant hosts causing different symptoms in each. Symptoms are therefore reflections of host response.

Most viruses spread through their host, producing *systemic infection*. Meristematic regions of roots and shoots may remain virus-free. In some cases, initial rapidly-produced *primary symptoms* may differ from those produced later. Where symptoms result in prompt death of invaded cells, thus preventing further spread of infection, plants are said to show a *hypersensitive reaction* and to be *hypersensitive*.

The most obvious symptoms are external and take the form of foliage colour changes or growth abnormalities. Internal symptoms are also produced but often these may be recognized only after careful optical or electron microscopy.

2.2 External symptoms

Virus infection may produce recognizable changes of leaves, stems and roots, flowers and fruit. The vernacular names of plant viruses and the diseases caused by them are based on the external symptoms of disease in a particular host. For example, beet yellows virus causes yellowing of sugar beet, whereas beet yellow stunt virus produces similar yellowing and more severe stunting of sugar beet (and also has a different host range).

2.2.1 *Leaf and foliage symptoms*

Generally, it is colour changes of leaves that produce the most noticeable symptoms; these are mainly *mosaic, mottle, yellowing, ringspot* and brown *local lesions.* Typically, the description 'mosaic' is applied to leaves with dark-green, light-green or yellow areas, often angular and bordered by veins. In the mosaic pattern, some groups of cells may be virus-free whilst discoloured cells contain virus (Reid and Matthews, 1966; Atkinson and

Figure 2.1　Abutilon mosaic virus producing typical mosaic symptoms in *Abutilon striatum.*

Matthews, 1967). Of the 222 individual viruses covered by the 1970–81 CMI/AAB descriptions (see bibliography), 83 are mosaic viruses, including those of apple, beet, celery, cowpea, cucumber, lettuce, maize, soybean and tomato. A commonly-propagated ornamental shrub, *Abutilon striatum* var. *thompsonii* (the spotted flowering maple) shows attractive variegation due to abutilon mosaic virus (Fig. 2.1). Although this virus is transmitted by whiteflies it does not appear to be a great danger to other plants.

In monocotyledons, mosaic areas on leaves become elongated parallel with the veins, resulting in *streak* or *stripe* mosaic symptoms, as found in wheat streak mosaic and barley stripe mosaic diseases. The term *mottle* is applied where variegated areas of the leaf are rounded rather than angular, as in carnation infected with carnation mottle virus.

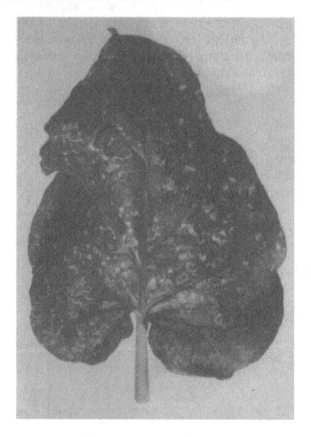

Figure 2.2   Ringspot symptoms produced by virus infection of tobacco.

Some viruses, e.g. beet yellows virus, induce yellowing or chlorosis of whole leaves or entire plants by lowering the chlorophyll concentration. Localized loss of chlorophyll and possible intensification of carotenoid pigments gives rise to *chlorotic spots* or *chlorotic local lesions* as in cucumber mosaic virus infection of *Nicotiana tabacum*.

Chlorotic rings, either single or concentrically arranged (Fig. 2.2), may be a major symptom caused by *ringspot* viruses, as in tobacco infected with tomato ringspot virus and raspberries and strawberries infected by raspberry ringspot virus.

Leaf colour changes may be associated with the veins of leaves, including *vein yellowing* (Fig. 2.3) or *chlorosis*, as produced by parsnip yellow fleck virus in parsnips. Vein mosaics and vein banding, where the variegated tissues are grouped along the main veins of leaves, are exemplified by red clover vein mosaic virus in legumes. In early stages of infection veins may be translucent, giving rise to the *vein clearing* symptom often confused with vein chlorosis. Vein clearing is a primary symptom found in lettuce plants with lettuce mosaic virus.

Most of the symptoms mentioned above arise as a result of disturbances of chloroplasts and decreased chlorophyll content in leaves. Leaves may also, however, show *reddening* or become purple, due to increased

Figure 2.3 Vein yellowing in *Pelargonium* containing leaf curl virus. Note also the colour of the younger leaves.

Figure 2.4   Local lesions produced by TMV in *Nicotiana glutinosa*.

anthocyanin production. This may occur in sugar beet with beet yellows virus. Oat leaves with barley yellow dwarf virus may become red or purple, and carrot leaves redden when suffering from carrot mottle dwarf disease, a condition caused by the simultaneous infection of carrots by a mottle and a red leaf virus.

Where viruses enter leaves killing surrounding cells, small brown necrotic *local lesions* are produced (Fig. 2.4). The word 'local' is used to emphasize that the lesion starts at the 'locus' or site of virus entry. These are primary symptoms. Holmes (1929) showed that counts of local lesions could be used to compare concentrations of TMV in *Nicotiana glutinosa* and *N. rustica*. For quantitative evaluation of virus preparations, it is useful to find a 'test plant' which reacts to give local lesions. Such lesions develop quickly, e.g. 2–3 days with tobacco necrosis virus (TNV) on French beans (*Phaseolus vulgaris*). Death of tissue at the locus of infection is a hypersensitive reaction and prevents spread of the virus through the plant. In some cases, however, virus may slowly invade other cells and lesions enlarge. In French beans, lesions produced by some strains of TNV often enlarge as virus spreads along the lateral veins, causing *veinal necrosis*.

Necrotic spots may vary from one plant species to another and can appear as necrotic ringspots. Necrosis, resulting in collapse of superficial tissue, is called *etch* (as in tobacco with tobacco etch virus), whereas death of epidermal cells causes *bronzing* of leaves in tomato with tomato spotted wilt virus. In very severe virus infection, necrosis of large parts or whole leaves may occur, as in pea early browning virus infections. Desiccation and withering of large portions of the leaf lamina may occur with tobacco necrosis infection of French beans.

Other symptoms shown by leaves include changes in texture, as in sugar beet infected with beet yellows where leaves are described as being yellow,

Figure 2.5 Leaf streaking and distortion in *Nerine* spp. with nerine virus.

Figure 2.6 Healthy and TMV infected leaves of French bean showing marked leaf distortion.

thick and brittle; these leaves splinter when crushed, and for this reason the disease was originally called 'crackly yellows'.

Virus-infected leaves are often smaller, misshapen, or blistered (Fig. 2.5). The lamina may be almost completely absent, as in *'shoe-string'* leaves of tomato and French beans infected with certain strains of TMV (Fig. 2.6). Uneven growth of the leaf lamina may result in *leaf curling* (Fig. 2.7) as in beet curly-top and also in potato leaf-roll diseases.

Outgrowths of the lamina surface, called *enations*, are a common symptom of pea enation mosaic virus infection of peas.

2.2.2 *Stem and root symptoms*

Stems may show mottle, streak or spot symptoms, similar to those of leaves, as a response to virus infection. Necrotic regions may appear on stems as the result of browning and death of epidermal and subepidermal cells. Necrosis of vascular tissue often results in wilting of leaves and stems, with subsequent withering of shoots. This occurs in peas with early browning virus and in gherkin plants (*Cucumis sativus*) with cucumber mosaic virus.

Potato tubers may show internal necrotic brown rings and spots (spraing) when infected with tobacco rattle virus or with potato mop top virus.

Shoots arising from potato tubers infected with potato virus X or Y may show irregular distribution and retarded anthocyanin production compared to healthy shoots (Martin, 1958). Anthocyanin production may also be disturbed in young apple twigs with apple mosaic virus (Baker and Campbell, 1966).

Other stem symptoms include swollen stems as in cacao plants (*Theobroma cacao*) infected with swollen shoot virus, and fasciation or flattened stems in grapevines infected with grapevine fan leaf virus. Stem tumours are produced on clover by wound tumour virus. Virus-induced rough bark diseases of woody plants cause flaking, cracking and necrosis of bark, together with excessive gum formation (Schneider, 1973).

Figure 2.7   Leaf curling in French bean due to bean common mosaic virus.

There are relatively few reports of root symptoms of virus infection, either because these are less easily observed or they are not very common. Possibly plants are less frequently examined for such symptoms. One particularly interesting disease, 'rhizomania' of sugar beet, has been a serious problem in Italy since the mid-1950s and is characterized by the abnormal proliferation of rootlets. A similar disease is found in Japan and France. This disease results from infection by beet necrotic yellow vein virus (BNYVV). A reduced number of adventitious roots is a feature of maize plants with rough dwarf virus. Root tumours are characteristic of wound tumour virus infections of white sweet clover (*Melilotus alba*). Cacao roots may be swollen with some strains of cacao swollen shoot virus, whereas tap roots of carrot and sugar beet are reduced in size when infected by carrot mottle virus and beet yellow viruses respectively; such reductions are of considerable economic importance. Woody tissue in roots may be discoloured in plants infected with alfalfa dwarf or tobacco yellow dwarf virus (Holmes, 1964).

The overall effects of virus on plants may be a reduction in growth, producing smaller stunted plants and a loss of yield of crop plants. There are a number of viruses that produce *stunt diseases*, e.g. beet yellow stunt and tomato bushy stunt. Other viruses cause similar dwarfing of their host, e.g. oat sterile dwarf, raspberry bushy dwarf, rice and satsuma dwarf viruses. In many instances, loss of photosynthetic efficiency reduces plant size because of yellows or mosaic type symptoms, whereas in other cases there may be some disturbance in the production or distribution of growth substance resulting in stunted bushy plants. Extreme stunting is to be found in shoots of potato plants with potato mop top virus.

Initial symptoms of virus disease may gradually increase in severity with progressive deterioration of the host. This effect is called *decline*, as in citrus decline caused by citrus tristeza virus.

2.2.3 Flower symptoms

Viruses may produce very striking changes in flower colour. In the 17th century Dutch tulip growers produced the much prized 'broken tulips' which are well known from Flemish paintings from as early as 1619. We now know that this *flower breaking* (Fig. 2.8) is due to infection by tulip breaking virus, and comes about by local loss or intensification of anthocyanin pigments giving rise to streaks and feathered patterns on petals. Similar flower breaking can be induced in anemone, petunia, stock,

Figure 2.8 Flower breaking symptom due to virus infection. (*A*) Stock with turnip mosaic virus. (*B*) Wallflowers with turnip mosaic virus. (*C*) Tulip with tulip breaking virus. Healthy plant on left, infected plant on right.

wallflower and zinnia by turnip mosaic virus. Narcissus yellow stripe virus may give flower colour changes in daffodil, jonquil and narcissus. Colour breaks in sweet william and carnation may arise following infection by carnation vein mottle virus. Chrysanthemums not only show colour breaking but flowers are also reduced in size and distorted in shape when infected with tomato aspermy virus.

Cocksfoot streak virus causes early flowering of the grass *Dactylis glomerata*, but on the other hand, cocksfoot mottle virus reduces flowering and also the yield of viable seeds.

Premature abscission of broad bean flowers is caused by bean leaf roll virus.

2.2.4 *Fruit and seed symptoms*

Fruits may show colour changes when the parent plant is invaded by virus. In plums, dark-coloured rings, lines or bands may appear on fruit with plum pox (Sharka) disease. Discoloration may arise from uneven ripening, and in red fruit, the pox mark may be present as discrete yellow mottling of the fruit. Similar mottling can be found on tomato fruit due to tomato spotted wilt virus, whereas the aucuba mosaic strain of TMV causes light and dark green blotching on tomato fruits. Cucumbers with mosaic virus become yellow-green mottled, but occasionally in late infection become greenish-white instead, hence the original American name 'white pickle' for this disease.

Bean common mosaic virus may cause brown necrotic areas and internal necrosis of the pods of legumes, and broad bean stain virus causes brown necrosis of the testa of broad beans.

Cucumbers infected by cucumber mosaic are not only mottled yellowish-green, but darker portions of the fruit produce wart-like projections and later the fruits are distorted and misshapen. Similar symptoms may be produced by this virus in melons and squash fruits. Irregular grooves and pits on plum fruits are also a characteristic symptom of plum pox disease. Passionfruits become woody with hard pericarps following infection with passionfruit woodiness virus and also with cucumber mosaic virus.

Tomato fruits from plants infected with aspermy virus are dwarfed, malformed, hollow and seedless. Seedless pods are produced by soybeans infected with soybean mosaic virus. Grain production is reduced in wheat with wheat streak mosaic virus.

2.3 Internal symptoms

Internal symptoms of virus diseases are those detected by light and electron microscopy, and will be considered under two headings, (*a*) anatomical and histological changes, and (*b*) cytological and ultra-structural changes. Virus-induced cell *inclusion bodies* will be considered separately in section 2.4.

2.3.1 *Anatomical and histological changes* (Figs. 2.9 and 2.10)

Plant pathogens, including viruses, may affect the size of plant organs and tissues by increasing cell numbers, a condition known as *hyperplasia*. Excessive growth due to enlargement of individual cells is termed *hypertrophy*. Only one term, *hypoplasia*, is used to denote underdevelopment of tissue due to reduced cell size or reduced cell numbers.

In leaves showing mosaic or other yellowing symptoms, the palisade cells may be smaller and contain fewer chloroplasts. The lamina in yellow areas is thinner, with less well differentiated smaller mesophyll cells and few or no intercellular spaces. Esau (1956) refers to this as an expression of hyperplasia. Dark green areas of leaves may develop normally, but buckling and blistering of the leaf lamina may arise when the growth of such areas is impeded by less actively growing cells in adjacent discoloured areas.

Extreme hypertrophy of the lamina gives rise to 'shoestring' leaves. Such leaves in tobacco plants infected with TMV result from a lack of dorsoventrality of the leaf primordia and an inhibition of meristem activity (Tepfer and Chessin, 1959).

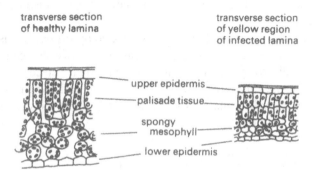

transverse section
of healthy lamina

transverse section
of yellow region
of infected lamina

upper epidermis

palisade tissue

spongy mesophyll

lower epidermis

Figure 2.9 Changes in leaf anatomy following infection by a virus giving mosaic symptoms. (Based partly on Esau, 1956).

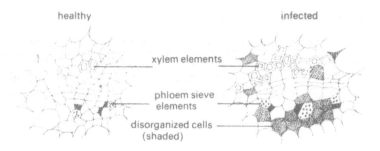

Figure 2.10 Diagrams of transverse sections of tobacco leaves to show xylem proliferation and cell enlargement (hypertrophy) resulting from virus infection.

Vein clearing appears to be due to enlargement (hypertrophy) (Fig. 2.10) of cells adjacent to veins where intercellular spaces become obliterated, few chloroplasts are produced and the tissue becomes translucent (Esau, 1956).

Anatomical changes are not limited to leaf lamina tissue, however. Phloem is often destroyed by invading viruses, resulting in phloem necrosis, as in potato leaf roll disease. Abnormal phloem development may occur, with hyperplasia of phloem parenchyma, in curly top virus disease of sugar beet. This phloem subsequently dies. The tumours of wound tumour virus result from abnormal meristematic activity of phloem parenchyma cells. Normal phloem sieve cells have sieve plates with deposits of callose ($\beta(1 \rightarrow 3)$-linked glucose polymer) in relatively small amounts lining the sieve pores. Some virus-diseased plants contain large quantities of callose that may block the sieve pores. Callose is thought to be deposited in mechanically damaged phloem sieve cells to prevent loss of valuable sucrose and other translocates. Presumably a similar defence mechanism comes into operation when sieve cells are disturbed during virus infection.

In woody plants, the cambium may be affected by viruses. Citrus tristeza virus, for example, causes localized areas of the vascular cambium to produce small groups of unorganized parenchyma instead of xylem or phloem. Because this parenchyma adheres to the bark, when the bark is peeled from the xylem pits are seen in the wood, giving the *stem pitting* or *wood pitting* symptom.

2.3.2 *Cytological and ultrastructural changes*

Viruses alter the gross form, arrangement and appearance of cells by disturbing their internal organization (Fig. 2.11). Such disturbances may

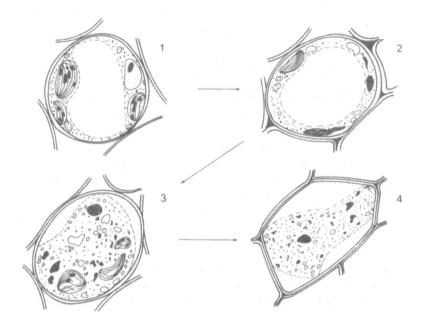

Figure 2.11   Changes in French bean leaf cell ultrastructure during local lesion production (based on Spencer and Kimmins, 1971). (1) Healthy leaf mesophyll cell; (2) Initial stages of infection; (3) Cell destruction in region near local lesion; (4) Complete loss of organelles in necrotic area of lesion.

be apparent when changes in structural features of organelles are involved. The main organelles influenced by viruses appear to be the *nucleus*, *chloroplasts* and *mitochondria*, with less well-documented changes in other cellular structures. In some instances, disintegration of organelles occurs as viruses multiply. In oats for example, oat mosaic virus causes deterioration of microbodies, chloroplasts, nuclei and mitochondria in that order, and the cells finally contain only remnants of these structures (Herbert and Panizo, 1975).

Nuclear changes. In some cases of virus infection the cell nuclei may stain less readily, as with eggplant mottled dwarf virus, and nucleoli may be swollen and distorted in beet following beet mosaic virus infection. These distortions of the nucleoli may result from the multiplication of virus in the nucleolus or nucleus since, with many viruses, virions can be observed by electron microscopy in the nucleoplasm. This is the case with barley stripe mosaic and pea enation mosaic. Most rhabdoviruses appear to replicate in the nucleus, and virus particles are often to be found aggregated in the perinuclear space or in tiers in the nucleoplasm.

Chloroplast changes. As mentioned earlier, the most common external symptoms of virus infection are leaf colour changes. In many cases, such colour changes involve yellowing of part or whole leaves and clearly reflect changes in the chloroplasts or their pigments. The presence of virus particles in chloroplasts is a rare phenomenon although virions of beet yellow virus have been found in plastids from beet (Cronshaw *et al.*, 1966) and tobacco infected with TMV (Esau and Cronshaw, 1967). More recently, rafts of alfalfa mosaic virus particles have been described in cytoplasmic invagination in chloroplasts (Hull *et al.*, 1970).

Chloroplast changes in virus-infected tissues have been particularly well documented by Esau (1968). Degeneration of chloroplasts varies from plant to plant but often follows a pattern of loss of thylakoid system, with disruption of stroma lamellae, accumulation of starch, and increase of lipid material in the form of osmophilic globules or plastoglobuli. Accumulation of starch and of osmophilic globules is not, however, specific to virus

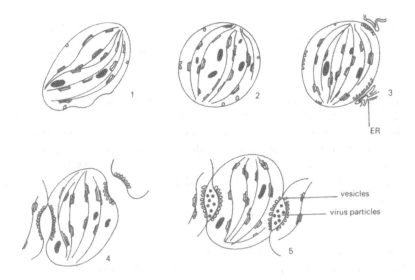

Figure 2.12 Sequence of changes in the chloroplasts of Chinese cabbage following infection with turnip yellow mosaic virus (based on Hatta and Matthews, 1974). (1) Chloroplast with scattered vesicles; (2) swollen chloroplast, vesicles scattered; (3) vesicles aggregated, endoplasmic reticulum (ER) associated with vesicles; (4) chloroplasts aggregate together in area of vesicles; (5) virus particles appear in space between chloroplasts.

infection and may occur with other pathological and physiological conditions. Swelling of chloroplasts may occur when sugar-cane cells are infected with sugar-cane mosaic virus, whereas chloroplasts in plants infected with cacao swollen shoot virus are smaller and flattened (Knight and Tinsley, 1958).

A characteristic feature of infection by tymoviruses is the formation of fibril-containing vesicles in chloroplasts (Fig. 2.12). The value of such vesicles for diagnosis of tymovirus infections is limited, however, since TMV (tobamovirus group) and beet yellows virus (closterovirus group) induce similar vesicles. Where vesicles are formed, they appear to arise from localized invaginations of the chloroplast limiting membrane (Hatta et al., 1973), and in early stages contain endoplasmic reticulum which provides an intimate connection between chloroplast and nucleus. These vesicular structures may also contain the enzymes and nucleic acid templates necessary for virus replication, particularly in the case of turnip yellow mosaic virus (TYMV).

The vesiculated chloroplasts of TYMV swell and clump together. Aggregates of chloroplasts are characteristic features of infection by other tymoviruses, and are also found with hydrangea ringspot virus (potexvirus group) and with barley stripe mosaic (hordeivirus group).

Mitochondrial changes. Mitochondria may also aggregate during virus infection of cells; such aggregates may be of either normal mitochondria or degenerate ones. With tobacco rattle virus, inclusion bodies commence development as small aggregates of almost normal mitochondria. As the aggregates enlarge, the mitochondria become filled with small vesicles containing fibrillar material resembling nucleic acid. In mitochondria of *Cucumis sativus* infected with cucumber mottle mosaic virus, vesicles are formed as soon as, or even before, virus production. Vesicles are formed by a single membrane and are located in the perimitochondrial spaces between outer and inner membranes and in the cristae.

Association between apparently normal mitochondria and some virions cause the organelles to aggregate together. Such clumping of mitochondria occurs with three rod-shaped viruses, tobacco rattle virus (TRV), with potato virus Y (PVY) and also henbane mosaic (HM) virus. Samples of TRV contain short particles (45–115 nm long) and long particles (190 nm). Long particles become attached to mitochondria by their ends and this causes clumping of the organelles. With PVY and HM, the virions appose sideways to the mitochondria causing aggregation. Mitochondria also aggregate in the presence of the isometric particles of broad bean wilt virus.

Other changes. Less well-documented changes occur in other organelles of infected cells. *Ribosomes* often become integrated into inclusion bodies, and may form fairly distinct masses as with carnation etched ring virus in *Saponaria vaccaria* (Larson, 1977). On the other hand, 70S chloroplast ribosomes disappear after symptoms of lettuce necrotic yellow virus appear (Francki and Randle, 1970). Particles of TMV may be found in the wide cisternal spaces of *endoplasmic reticulum* (ER) (Esau, 1968). ER is often enclosed in virus inclusion bodies. Virus particles may accumulate in the *cytoplasm* or may be closely associated with each of the organelles mentioned above. Viruses such as TMV and beet yellows may also accumulate in the vacuole (Esau, 1968) and pea seed-borne mosaic virus accumulates on the membranes of the vacuole (tonoplast).

Potato leaf-roll virus (PLRV) and potato mop top virus infections are associated with the appearance of tubules in cells. In recent work on PLRV (Shepardson and McCrum, 1980) it has been suggested, however, that such tubules are not absolutely specific to the presence of PLRV infection in PLRV-free cells, and may simply represent aggregation of somewhat non-typical plant microtubules.

Virus infection may result in proliferation of the plasmalemma to produce groups of vesicles making *paramural bodies*. Such bodies are frequently associated with bean pod mottle virus. This virus has also been reported to be embedded in cell walls within tubules.

Cell walls may be modified either by the deposition of callose or by the production of finger-like projections of the wall. Callose deposition is induced by some elongated viruses in local lesion reactions and may help to restrict virus movement. Finger-like projections of cell walls are associated with plasmodesmata and occur in both local lesion and systemic infections.

2.4 Inclusion bodies

Inclusion bodies have been defined as 'intracellular structures produced *de novo* as a result of virus infection. These structures may contain virus particles, virus-related materials, or ordinary cell constituents in a normal or degenerate condition, either singly or more often in various proportions' (Martelli and Russo, 1977). Inclusion bodies vary in form from amorphous structures to well-defined crystalline structures; some are large enough to be seen in the light microscope, others can only be detected with the electron microscope. For convenience, plant virus induced inclusions will be considered under two headings: (*a*) nuclear inclusions, and (*b*) cytoplasmic inclusions.

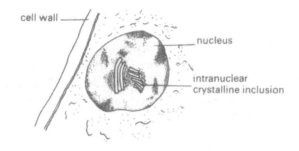

Figure 2.13 Intranuclear inclusions induced by virus infection (after Rubios-Huertos, in Gibbs and Harrison, 1976).

2.4.1 *Nuclear inclusions* (Fig. 2.13)

These may occur in the nucleoplasm, in the nucleolus or between the membranes of the nuclear envelope (perinuclear space).

Inclusions in the nucleoplasm (the body of the nucleus), are generally crystalline in structure, although rounded and amorphous types have been recorded from cotton and from celery. Fibrous inclusions, visible only in the electron microscope as aggregates of tubular units or of tubular viruses, have also been reported, as with beet yellow virus and TMV. Intranuclear membrane inclusions may be present in cells infected with cowpea mosaic virus. These are discrete accumulations of vesicles which form compact structures in the nucleus and are probably derived from the nuclear envelope (Langenberg and Schroeder, 1975).

Crystalline inclusions in the nucleoplasm are of two main types: (*a*) protein crystals, and (*b*) crystalline arrays of virus particles. The nuclear protein crystals of tobacco etch virus resemble truncated four-sided pyramids, and may appear also in the cytoplasm, having originated in the nucleus. The protein of these crystals appears to be different from virus protein. Several isometric viruses also produce distinct crystalline inclusions, and these may be composed of virus particles, although those of turnip yellow mosaic virus appear to be composed of empty capsids (Hatta, 1976).

Nucleolus-related inclusions may assume an amorphous or crystalline organisation. In some cells, such as those of *Gomphrena globosa* and *Chenopodium quinoa* infected with beet mosaic virus, accumulations of intensely electron-opaque protein can be observed as dense masses of amorphous or finely granular material localized at the periphery of the nucleolus. Such accumulated material constitutes 'satellite' bodies, and these enlarge as infection progresses.

Nucleolar inclusions may also be crystalline, particularly in plants infected with bean yellow mosaic virus (BYMV). Such inclusions are composed of protein, are 0.2–3 μm in diameter and may distort the shape of the nucleolus. Similar crystalline nucleolus-related bodies have been described for potyviruses.

The accumulation of viruses or virus-induced material between the membranes of the nuclear envelope gives rise to *perinuclear inclusions*. Such inclusions are often transitory and may represent material being moved from the nucleus to the cytoplasm or vice versa.

Perinuclear inclusions composed of virus particles are commonly found after infection by rhabdoviruses. In some cases, accumulation of virions may be large enough to become visible in the light microscope. Often, however, virus particles simply line the perinuclear space, being budded from the inner lamella of the nuclear membrane, as with eggplant mottled dwarf virus and maize mosaic. Vesicles produced from the inner membrane lamina make up the perinuclear inclusions induced by pea enation mosaic virus.

2.4.2   Cytoplasm inclusions

Virus particles may be scattered in the cytoplasm of infected cells, but are often difficult to locate because of their low concentration. In systemically infected leaves showing marked external symptoms, viruses may multiply at high rates, producing diagnostic inclusion bodies which vary enormously in size, shape, location, composition and organization; they include

  (a) amorphous inclusions
  (b) fibrous, banded and paracrystalline bodies
  (c) crystalline inclusions
  (d) pinwheel and laminated inclusions
  (e) viroplasmas and complex structures.

Amorphous inclusions (Fig. 2.14). In 1903, Ivanowsky described crystalline and amorphous structure in tobacco cells from TMV-infected plants. Attention was paid to the amorphous bodies, since they were thought to be amoeboid. Goldstein (1924) named these structures *X-bodies* because of their unknown nature. Sheffield (1931–34) showed that the X-bodies stained strongly for protein and were more stable than crystals, and also isolated whole inclusions and showed them to be highly infectious,

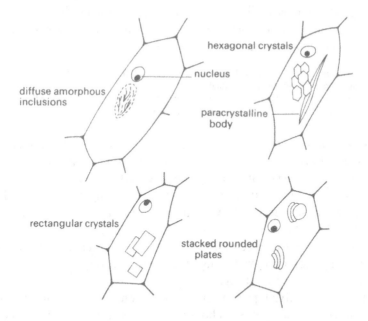

Figure 2.14 Various cytoplasmic inclusions induced by plant viruses.

containing virus particles plus other material. Electron microscope studies have shown that X-bodies contain accumulations of golgi bodies, ribosomes, endoplasmic reticulum and mitochondria together with tubules or filaments of protein, possibly concerned with virus replication (Shalla, 1964).

TMV X-bodies may be 5–30 µm diameter, the larger bodies being associated with large cells and vice versa. The bodies are granular in appearance, containing vacuoles, and occur in the cytoplasm close to the nucleus. Matsui (1959) divides X-bodies into two groups, those elliptical in outline and almost entirely composed of dense granules of varying sizes, and those spherical in outline, consisting of a narrow peripheral zone of dense granules surrounding a large central vacuole. Warmke (1969) has proposed that the term 'X-body' be used only for amorphous inclusions associated with TMV infection, the term *amorphous inclusion* or *amorphous inclusion body* being used for non-crystalline inclusions similar to X-bodies found in association with other virus infections. Amorphous inclusions seem to be associated not only with members of the tobamovirus group but also with potyviruses such as bean yellow mosaic and beet mosaic viruses.

Large amorphous bodies composed of proteinaceous sheets are consistently found associated with potato virus X infection. These are called *laminated inclusions* and may be of three types:

    (*a*) those composed of protein sheets mostly virus-free but heavily studded on both sides with 14 nm diameter beads (beaded sheets);
    (*b*) beaded sheets with virus particles interspersed between the layers, and thirdly,
    (*c*) smooth surfaced sheets with numerous virions between them.

Fibrous, banded and paracrystalline bodies (Figs. 2.14, 2.15). Pure or nearly pure aggregates of virus may accumulate in cell cytoplasm in various ways. Particles may aggregate at random with no orderly orientation, giving rise to *fibrous inclusions*. If particles align themselves side by side and/or end to end in a two-dimensional way, then *paracrystalline* structures are produced. True *crystalline inclusions* arise by the orderly accumulation of virus particles into a three-dimensional lattice. Fibrous inclusions have been described for a very wide range of elongated viruses in the potexvirus, carlavirus and potyvirus groups.

Another distinctive type of inclusion body contains virus particles and has a banded appearance. Such bodies are produced by PVX. Viruses can accumulate by an orderly alignment of particles in layers stacked one upon the other. Banding arises because each layer is not in contact with its neighbouring stack, but is separated by thin cytoplasmic strands, often containing ribosomes, presumably trapped there during formation. Such *banded bodies* may reach a considerable size and be visible in the light microscope, e.g. banded inclusions from the phloem of *Beta vulgaris* infected with beet yellows virus are about 50 μm long and 3–4 μm wide.

TMV-infected cells may contain linear aggregates of virions appearing as long fibrous structures when seen in the light microscope. These

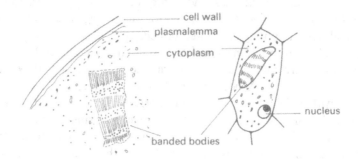

Figure 2.15 Cytoplasmic banded bodies in parenchyma cells.

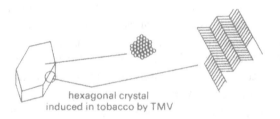

hexagonal crystal
induced in tobacco by TMV

Figure 2.16 Diagram to show the electron microscope appearance of crystalline inclusions in sections.

structures may extend the whole length of the cell as straight needles, refractive spindle-shaped bodies or even as long looped bodies. These virus aggregates do not have the three-dimensional ordered structure of crystals and are therefore *paracrystalline bodies*. TMV particles may be arranged in stacks of parallel elements orientated across the long axis of the aggregate. Where each layer is placed over the next at an angle, *angled-layer aggregates* are formed.

Paracrystalline bodies are produced by most rhabdoviruses and by isometric viruses such as comovirus, nepoviruses and plant reoviruses.

Figure 2.17 Section through crystalline body composed of TMV particles.

Crystalline inclusions. Crystalline inclusions (Figs. 2.16, 2.17) are produced by both elongated and isometric viruses. Ivanowski (1903) discovered, by light microscopy, crystalline structures in tobacco leaf cells infected with tobacco mosaic virus. Crystalline inclusions occur in epidermal and hair cells. Recently, studies of these crystals using the electron microscope have produced evidence that they are composed largely of virus rods. In thin sections, TMV crystals appear as stacked layers of closely aligned parallel arrays of virus particles. Successive rows of viruses may be slightly tilted, forming angles between rows, and giving 'herringbone' patterns of particles.

Crystalline inclusions arise by the gradual accumulation in the cytoplasm of virions to form small aggregates of parallel rods with ends aligned. Such aggregates enlarge, initially becoming monolayers, which then stack together to give true three-dimensional crystalline structures.

Most groups of isometric viruses form true crystals, varying in size from those detectable only by the electron microscope to those 15–20 µm in diameter. X-ray diffraction studies of tomato bushy stunt, turnip yellow mosaic and tobacco ringspot viruses indicate that they have cubic symmetry, since their crystal lattice is centred on a cubic unit cell, composed of a central particle and one at each corner of a cube around

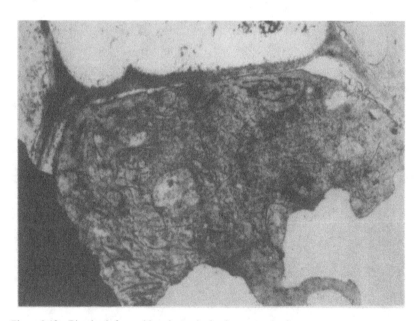

Figure 2.18   Pinwheels formed in tobacco by henbane mosaic virus.

that particle. Studies of southern bean mosaic virus suggest that the arrangement is based on a rhombohedral or orthorhombic structure (Johnson *et al.*, 1974; Akimoto *et al.*, 1975).

Pinwheels and laminated inclusions. Some viruses produce complex three-dimensional proteinaceous structures (Figs. 2.18, 2.19) which in section appear as a group of curved membranous arms diverging from a central core. This central core is tubular and the radiating laminate arms taper from one end to the other, so that in some sections the arms are short and in others long. These are called *pinwheels.* Some workers consider that these structures are the central portion of *cylindrical inclusions* and may originate in or near plasmodesmata. As well as appearing in cross section as pinwheels, cylindrical inclusions may appear as *scrolls* resulting from the extension and rolling up of pinwheel arms. Associated also with cylindrical inclusions are *laminated aggregates* which are composed of several thin, flat plates in parallel orientation, often near or attached to pinwheel arms. Pinwheels, scrolls and laminated aggregates are characteristic of the potyvirus group. Edwardson (1974) separates potyviruses into three subgroups according to the presence of the following structures

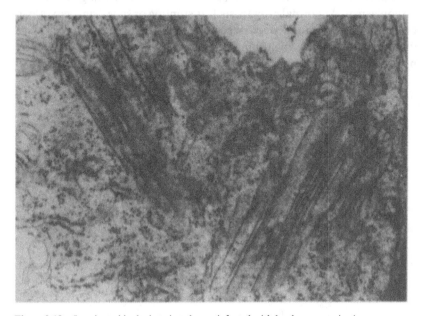

Figure 2.19   Laminated inclusions in tobacco infected with henbane mosaic virus.

attached to the central portion of the pinwheel inclusion: subdivision I—scrolls or tubes; subdivision II—laminated aggregates, and subdivision III—scrolls and laminated aggregates.

How useful these inclusions are in virus classification remains to be seen, since the type and number of inclusions may vary with degree of infection (Andrews and Shalla, 1974), and with the technique and type of fixative used in preparing material for examination in the electron microscope (Langenberg, 1979).

Viroplasms. Finally, many virus-infected cells contain collections of finely textured, electron-dense materials in which viral products are formed and/ or virus assembly takes place. These aggregates are termed *viroplasm matrices* or *viroplasms.* These viroplasms in advanced stages of infection are compact bodies lacking membranes and are consistently associated with infections by reoviruses and members of the caulimovirus group.

2.5   Virus-like symptoms with other causes

Beware! Not all virus-like external symptoms are caused by viruses. Yellowing of leaves may be due to mineral deficiency, senescence and physiological and genetic causes, as well as to other pathogens such as mycoplasmas. Abnormal growth, such as stunting or shoestring leaves, can be caused by hormone weedkillers. Bacterial pathogens may produce tumours and cankers. Inclusion bodies may not be produced exclusively by virus; for example, nuclear inclusions are often a common feature of some plant cells.

CHAPTER THREE

TRANSMISSION OF PLANT VIRUSES

Plant viruses are obligate parasites, often causing the death of their host, so it is necessary for them to spread from plant to plant and to be introduced into living cells. This spread or transmission will be considered under the following headings:

(1) mechanical transmission
(2) vegetative, graft and dodder transmission
(3) transmission by pollen seeds
(4) insect and mite transmission
(5) nematode and fungal transmission.

Organisms involved in the transport of virus from one plant to another are called *vectors*. Transmission by vectors involves some transitory biological interaction between vector and virus, but in many cases a specific vector will interact only with a particular virus.

3.1 Mechanical transmission

This may be brought about (*a*) by plants making contact, (*b*) by the action of animals, but with no biological interaction or specificity being shown, and (*c*) by action of humans, either accidentally or deliberately for experimental purposes.

(*a*) Mechanical transmission by plant contact can occur in the case of systemically infected plants, where virus invades most of the tissues and is present in epidermal cells including leaf and stem hairs. As a result of plants touching, hairs and epidermal cells may be broken and virus liberated into similarly damaged cells on healthy plants, thereby bringing about transmission. It seems likely that a large number of virus particles need to be transferred in this way before infection results, but such

41

transmission does occur and results in intensification of disease within closely planted crops.

(b) Animals, such as rabbits, foxes and dogs, may transfer virus by first rubbing against infected plants and then entering healthy crops. Virus picked up on the animal's coat is then transferred to healthy crops, mainly through minor damage or injury to plants. Broadbent (1963) has shown that birds may transfer TMV by this method.

Some insects, such as large beetles and grasshoppers, are thought to transfer viruses such as turnip mosaic, TMV and PVX by purely mechanical means, virus being picked up on their mouthparts during feeding and then inoculated into plants on subsequent feeding. In these cases there is no special interaction between the insect and the virus and no specificity of insect species with type of virus.

(c) Mechanical transfer of virus between plants can be brought about by man. For example, PVX can survive on clothing, tools and mechanical devices used in agriculture and horticulture and these can therefore act as sources of virus when used in otherwise virus-free crops. Broadbent (1963) tells how a worker, in overalls worn for two days in a TMV-infected crop, worked between two rows of potted tomato plants for ten minutes with the result that 19 of the 20 plants became infected. Clothing contaminated with TMV may remain infective for several months.

Virus may also be transferred from infected to healthy plants on cutting knives when pruning or harvesting produce. Viruses such as TMV and tomato mosaic may be spread too when pricking out seedlings or harvesting fruit and flowers by hand. Fortunately, however, not all viruses are mechanically transmissible.

3.1.1. Experimental mechanical transmission

In order to establish the viral etiology of a disease or to identify a known virus disease, extracts of the diseased plants are rubbed on to the leaves of test plants. A wide range of test plants is used, including species of tobacco, legumes, *Chenopodium*, cucumber, wheat and barley (see Table 7.1). Entry of the suspected virus into living cells of the test plants is facilitated by including an abrasive, such as carborundum powder, in the inoculum. The extract of sap is applied to the leaf using a finger (Fig. 3.1), or on a brush or pad of gauze. The success of inoculation by this method depends on numerous factors, including virus purity and stability, pH and ionic strength of inoculum, type, age and physiological condition of the host, and also pre- and post-inoculation conditions. Why many viruses are not

Figure 3.1 Inoculating leaves of French bean with a virus. The leaves are dusted with carborundum powder and the inoculum applied from the watchglass.

readily transmitted mechanically in spite of careful consideration of these factors is not understood; possible explanations include inactivation of virus during extraction, the presence of inhibitor compounds in extracts or in epidermal cells of prospective hosts, or the requirement for virus to enter specific cells beneath the epidermis.

3.2   Vegetative, graft and dodder transmission

3.2.1   *Vegetative transmission*

In a systemically infected plant, virus is to be found in nearly all organs and tissues of the plants, and it therefore follows that parts used for vegetative propagation will contain virus and so give rise to infected plants. It is for this reason that growers are advised not to save potato tubers for seed purposes. Similarly, care should be taken not to propagate strawberries, carnations, chrysanthemums or dahlia for example, from diseased or suspect plants.

The apical meristems of systemically infected plants are, however, virus-free, and with skill and care it is possible under the right aseptic conditions to isolate the tip and from it grow whole plants free of virus (see Fig. 6.1).

3.2.2 *Graft transmission*

Grafting (Fig. 3.2) consists of joining an isolated shoot (scion) or bud of one plant to the rooted part (root stock) of another plant. The union of the two effectively produces one plant. Commercially, grafting is used to propagate choice plants that are otherwise difficult to root, and to give rapid production of these plants; furthermore many plants yield better growth if on wild rootstock rather than on their own roots. Where one part of the union contains the virus, the plant as a whole may become infected.

Graft transmission of viruses has been known for a long time. Although the cause of flower colour breaking was not understood, Blagrave in 1675 (see McKay and Warner, 1933) gave exact details of transmission of the condition by grafting bulbs together. Similarly, Cane in 1720 reported that he had transmitted the mottled or variegated condition of jasmine to white non-mottled plants by grafting. Graft transmission of disease has been used as a diagnostic test and interpreted as indicating a viral etiology for

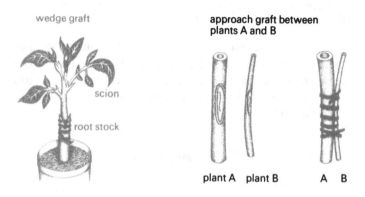

Figure 3.2 Two types of grafting used to transfer virus from one plant to another.

disease, but the more recent findings of mycoplasma-like organisms in plants make this test less reliable.

Transmission by 'approach grafting' allows virus to move from very dissimilar species. The approach graft union is by callus cells produced at the contact surfaces, and differs from true graft unions where vascular tissues unite by mutual activity of their vascular cambia. Such intimate unions are restricted to closely related plants.

Transmission by grafting may result in the production of different symptoms to those found following sap inoculation. In *Nicotiana glutinosa*, TMV induces local lesions and there is no systemic virus spread when it is mechanically sap-inoculated into healthy plants. Similar plants grafted with tomato, or tobacco systemically infected with TMV, die as the virus spreads systemically, producing necrosis of leaves and buds. This effect probably comes about because virus is introduced by the graft into vascular tissue; alternatively, more virus may be introduced into plants via graft unions than by mechanical or other methods. A particularly important disease of citrus fruit is that caused by citrus tristeza virus. This virus, which infects oranges, grapefruit and limes, causes little loss of many commercial sweet orange varieties or of sour orange trees on their own rootstock. Unfortunately, however, sweet orange, grafted for better productivity onto sour orange rootstock, is killed by the tristeza virus. This forms the basis for testing for this virus. Thus, citrus tristeza virus can be detected in symptomless *Citrus sinensis* (sweet orange) plants by grafting onto *C. aurantium* (sour orange) when the rootstock suddenly wilts, declines and dies. Similarly, plum pox was originally detected by grafting suspected material on to peaches. However, this rather lengthy, not altogether reliable test has now been replaced by serological testing.

Grafting, widely used commercially for plant propagation, may assist in the spread of virus disease, particularly in apple, pear, plum, grapevine, and citrus, and ornamental trees and shrubs such as roses.

Graft transmission is for the most part an artificial method of virus transmission but natural grafts between plants have been implicated in the transmission of carnation mosaic virus (Thomas and Baker, 1952) and apple mosaic (Hunter *et al.*, 1958).

3.2.3 *Dodder transmission*

A method of transmission which is experimentally useful and which can be considered an extension of graft transmission is that using dodder (*Cuscuta*) (Fig. 3.3). Dodder plants are parasitic members of the family

Figure 3.3 Dodder transmission of viruses. Picture shows dodder, a thread-like plant, connecting two different species together.

Convolvulaceae. There are many different *Cuscuta* species; two, *C. europaea* and *C. epithymum*, are native to Great Britain. Dodders arise from seeds that often only germinate in the presence of specific hosts. The stems, which are leafless and lack chlorophyll, elongate and wind around suitable plants, sending out haustoria which penetrate to the vascular tissue of the host at points of contact. Nutrients are derived from the host by these root-like haustoria, allowing the parasite to grow, branch and eventually flower. Branches from the initial dodder stem entangle other plants and parasitize them. Virus can pass along the connecting dodder stems between plants. Whether this pathway functions in nature is not documented, but it can be used to transmit viruses experimentally that are not easily transmitted by other means. Viruses such as cucumber mosaic and tobacco rattle replicate in the dodder and are more efficiently transferred than, say, TMV which is merely carried on the dodder without infecting the parasite. Dodder itself may be infected with dodder latent mosaic virus and this may produce symptoms in plants that are parasitized by the infected dodder.

Table 3.1 Some examples of seed- and pollen-transmitted viruses.

Virus	Host	Percentage transmission	Pollen transmission	Alternative transmission
Arabis mosaic	Capsella-bursa pastoris	6–33	–	Nematodes
Barley stripe mosaic	Barley	15–100	+	Unknown
Bean yellow mosaic	Peas	10–30	–	Aphids
Cucumber mosaic	Cowpea	4–28	–	Aphids
Grape-vine fan leaf	Chenopodium quihon	3	–	Nematodes
Prune dwarf	Cherry	9	+	Unknown
Raspberry ringspot	Strawberry	50	–	Nematodes
Soybean mosaic	Soybean	0–68	–	Aphids
Tobacco rattle	Capsella-bursa pastoris	2	–	Nematodes
Tomato ringspot	Soybean	76	–	Nematodes

3.3 Seed and pollen transmission

Seed transmission of viruses comes about by means of virus particles within the seed tissues, and results in the production of virus-infected seedlings. Virus may enter the seed from either one or both parental plants.

When virus enters seeds via pollen originating from infected plants, the virus is best described as *pollen-borne*. When virus-bearing pollen brings about the infection of the ovule-bearing plants, the virus is *pollen-transmitted*. The distinction between pollen-borne and pollen-transmitted virus is not often clearly stated in the literature.

A wide range of viruses is seed-transmitted, although often the percentage of seeds carrying virus is small (Table 3.1). The extent of seed transmission may depend on many factors, including type and strain of virus, type, strain and physiological condition of host plant at time of infection, and seed formation. Failure to transmit virus via the seed may be attributed either to failure of virus to enter developing seeds, or to inactivation of virus within seeds. To produce virus-infected seedlings, the embryo needs to be invaded, and often virus may be found only in non-embryo portions of the seed.

Seed-transmitted viruses have other means of spreading, often via nematodes. Seed transmission, however, provides an effective means of widespread dispersal of the pathogen, as well as allowing survival of virus between growing seasons and during times adverse to plant growth.

3.4 Transmission by arthropod vectors

Insects constitute the largest group of arthropod vectors of plant viruses. Large chewing insects may transmit viruses in a mechanical way as described earlier (3.1); however, transmission by most insects and by Acarina (mites) is a more complex event or series of events.

Table 3.2 shows the most important groups of insects concerned with plant virus transmission, and each will be considered in turn.

3.4.1 Aphids

These are the largest group of insect vectors and are probably the most important. In order to understand their role as vectors, a knowledge of their biology is essential.

Aphids (Figs. 3.4, 3.5), commonly known as greenflies or blackflies, are an extremely successful group. Individuals are small (2–3 mm only) and often inconspicuous, although their powers of reproduction are such that 0.4 hectares (one acre) of plants may contain 2000 million individuals and roots may support a further 260 million (Dixon, 1973). Many aphids are

Table 3.2 Insect vectors of plant viruses.

Order   Hemiptera—true bugs
   Sub-order   Heteroptera
      Family   Piesmidae—beet bugs
   Sub-order   Homoptera
     Series   Sternorrhyncha
      Family   Aphididae—aphids
      Family   Coccidae—mealy bugs
      Family   Aleyrodidae—whiteflies
     Series   Auchenorrhyncha—leaf and plant hoppers

Order   Thysanoptera—thrips
Order   Coleoptera—beetles
      Family   Chrysomelidae—leaf beetles
      Family   Curculionidae—weevils

Plant viruses may also be transmitted in a mechanical way by adults and juvenile stages of Orthoptera—grasshoppers and crickets, Dermaptera—earwigs, Lepidoptera—butterflies and moths, and Diptera—true flies.

Figure 3.4 A group of adult and young aphids feeding.

Figure 3.5 Aphids showing extended labium (proboscis) and stylets.

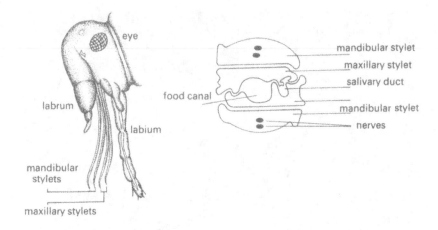

Figure 3.6　Details of aphid mouthparts. Left, external view of head and mouthparts; right, transverse section through stylets.

agricultural pests in their own right, since they suck sap from the phloem of plants. It is because of this mode of feeding that aphids transmit viruses. In these sucking insects the mouthparts are modified for sap feeding. The mandibles and maxillae, which are jaw-like structures in chewing insects, form two pairs of long bristle-like *stylets*. The maxillary stylets come together, their inner faces forming a food canal and a smaller salivary duct. On the outside are the two mandibular stylets (Fig. 3.6). The stylets run together in the anterior groove of the proboscis. The proboscis is held along the underside of the thorax in the resting position (Fig. 3.7). Feeding consists of bringing the proboscis forward and placing its tip against the

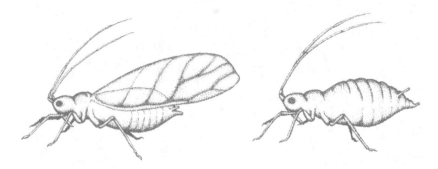

Figure 3.7　Alatae and apterae: winged and wingless forms of aphids.

plant surface. The stylets then penetrate the surface by alternate protraction of the mandibular and then the maxillary stylets. The mandibular stylet tapers to about $0.04\,\mu m$ or less in diameter, so that pressure applied at the base of the stylets results in enormous pressure at the tip. The stylets penetrate between cells rather than through them. Penetration is assisted by secretions, from the salivary glands, containing cellulase and pectinase enzymes. Initial probes are made into or between epidermal cells to test the palatability of the tissue, and during these probes the saliva sets as a gel which is left behind as a sheath. When the insect is satisfied with the food plant, the stylets are pushed through the tissues to the phloem, an operation taking a few minutes or sometimes several hours to achieve. Sap enters the aphid under pressure, although the insects can control the flow into their bodies by a muscular pump at the entrance to the pharynx. Saliva is also produced during feeding.

Most aphids are *monophagous*, living on a single type of host plant, but those of agricultural importance are host-alternating, or *polyphagous*. Thus the polyphagous peach-potato aphid *Myzus persicae* appears in spring on peach, having overwintered there as eggs. In summer it moves to a variety of secondary hosts including potato, tomato and sugar beet. Aphids are parthenogenic—producing young without fertilization—and also viviparous, the young being born alive. Sexual reproduction with the production of eggs also occurs.

As well as sexual and asexual forms aphids occur as winged (alatae) and wingless (apterae) individuals (Fig. 3.7). Overcrowding seems to trigger the change from wingless to winged forms. Winged aphids can fly from one host to another. Flight is usually over short distances at speeds of 1.6–$3.2\,km\,h^{-1}$, and convection currents during the day carry aphids into the air. Since the air moves faster than $3\,km\,h^{-1}$, the aphids have little control over their destination and can be carried great distances. After landing on plants, aphids make exploratory probes to test the palatability of the leaf and either settle down to feed or withdraw their stylets and make preparations for take-off.

In experiments and studies of insect transmission of viruses the following terms are often employed. The *acquisition feed* or *acquisition feeding* period is the time for which a virus-free vector actually feeds on a virus infected plant. This differs from the *acquisition access period* which is the time for which a vector is allowed upon a source of virus. Having acquired virus there may be a waiting period or *latent period* before the virus can be transmitted. Having acquired virus the vector may be placed on virus-free plants for a period of time known as the *inoculation access*

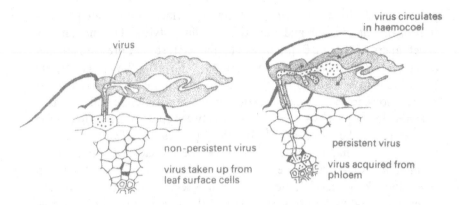

Figure 3.8   Virus acquisition by aphids.

period during which it could feed and transmit virus. The actual period of feeding is called the *inoculation feeding period*. The minimum period of time that a vector needs for the acquisition and subsequent inoculation of a virus-free plant is the *transmission time* or *transmission threshold period*.

Plant virus-aphid vector relationships are classified according to the length of time the virus continues to be transmitted by the insect; thus *non-persistent viruses* are retained for a few minutes or at most a few hours; *semi-persistent viruses* for several hours or a few days and *persistent viruses* for many days and often for the lifetime of the aphids. These terms are also used in connection with other animal vectors. Each of these categories of virus has distinctive transmission characteristics which demonstrate clearly that transmission is too complex to be simply a mechanical process.

Non-persistent viruses (Fig. 3.8). These have the following transmission characteristics.

(1) They are able to be mechanically transmitted.
(2) They are acquired by aphids with short acquisition feeds.
(3) The transmission threshold time is short.
(4) Transmission lasts only for a short time.
(5) Short acquisition feeds are better than extended feeding.
(6) Prolonged inoculation feeding reduces efficiency of transmission.
(7) Preliminary fasting increases capacity of aphid to transmit.
(8) Aphids' infectivity is lost upon moulting (ecdysis).
(9) Aphids show specificity.

These characteristics are interpreted as showing that virus is acquired from superficial cells of the plant during short probes of the tissue.

Acquisition probes may be as short as 5 seconds, indicating that virus is acquired from epidermal cells—probes of 60 seconds or more are required for stylets to penetrate beyond the epidermis (Roberts, 1940; Bradley, 1952).

The rapid acquisition and rapid loss of these viruses by aphid mouthparts at first looks suspiciously like mere mouthpart contamination and mechanical transmission. The term *stylet-borne* viruses is used by some virologists for the non-persistent virus because of the apparent contamination of stylet-tips and the passage of virus from there.

The loss of efficiency of transmission following long acquisition feeds may mean that less virus is available from deeper tissues, although Nambra (1962) has shown that cucumber mosaic virus is present in mesophyll cells and available to aphids. Bradley (1952, 1961) and Sylvester (1954, 1962) have suggested that prolonged probing removes more virus from the outside of the stylets than brief probes because the salivary sheath is longer and has hardened.

Watson and Roberts (1939, 1940), pioneers in studies of aphid transmission, suggested that the effects of pre- and post-access feeding in reducing transmission could be explained in terms of virus inhibitors produced during feeding. Such inhibitors have not been satisfactorily demonstrated. Another possible explanation put forward by Bradley (1952), following observations of feeding, is that starved aphids make brief probes of the type associated with successful transmission, whereas well-fed aphids make prolonged probes of the type ineffective for transmission. Furthermore, during prolonged feeding the stylets remain unsheathed by the labium (proboscis), and not being rigid could not penetrate new feeding sites. Delay in transmission might well represent the time required for stylets to be re-ensheathed prior to satisfactory probing and virus transmission.

The question remains—is virus located on the tip of stylets? Virus has not been detected on stylets. By washing stylets of *M. persicae* with water, treating them with formalin, or irradiating with ultra-violet (UV) light, virus transmission after access to PVY could be reduced. Although this would indicate that virus had been washed from the stylets or had been inactivated, there was also a change in the probing behaviour of the aphids. Furthermore, irradiation of stylets before access to virus also prevented subsequent acquisitions and transmission.

Some workers consider that virus is carried in the fore-gut rather than on the stylets. Virus is unlikely to be regurgitated from the aphid stomach or get beyond the fore-gut because the oesophageal valve prevents regurgitation and the stomach is retained during moulting, whereas the fore-gut is lost, corresponding to the loss of virus at ecdysis. Using plants labelled with P^{32}, Garrett (1971) has shown that aphids carried less label after 6–8 mins probe than insects allowed to probe for 3–5 mins only, suggesting that a large proportion of the label is returned to the labelled plant. The volume of sap represented by the label was too large to be accommodated in the stylets and it seems most likely to have been held in the fore-gut of the insects. Using *M. persicae* and *Hyperomyzus lactucae* it was shown that cucumber mosaic virus and P^{32} were transmitted in similar ways.

As a result of the uncertainty of the relationships between non-persistent viruses and the stylets of aphids many workers doubt the justification of using the term 'stylet-borne'.

Persistent viruses (Fig. 3.8). These viruses, after being ingested by aphids, are not immediately transmitted but there is a *latent period*, after which the aphid remains infective for a long period and in some cases for the remainder of its life. Infectivity is not lost upon moulting.

Persistent viruses are not easily mechanically transmitted and are acquired after long feeding times. These features arise because the viruses are to be found mainly in deep-seated tissues, mainly the phloem. Aphids may be able to acquire sufficient virus to become infective after 5–15 mins feeding, but the efficiency of transmission increases with increased access to the source of virus. Some virus may pass completely through the insect and be excreted (Richardson and Sylvester, 1965), but to be transmitted, virus particles have to pass through the gut wall, through the blood-filled body cavity (haemocoel) and enter the salivary glands. Such viruses are for this reason called *circulative*. Uptake of viruses is probably non-specific, the gut wall being the first barrier and site of specificity of entry of virus into the haemocoel and body tissues. The latent period represents the time taken for the virus to circulate around the insect. Pea enation mosaic virus for example is transmitted by *Acyrthosiphon pisum* and *Myzus persicae*. Nymphs can acquire virus in 15 mins, adults in 1–2 hrs, and the latent period of 4–70 hrs is temperature-dependent. After the latent period the virus can be retained by moulting individuals for up to 30 days, the virus being present in the lining of the mid-gut. The rate of transmission of these circulative viruses declines as virions are lost from the insects.

Circulative-propagative viruses. Here the story of the aphid transmission takes a more exciting turn because now the vectors act as host for the viruses they are transporting.

Viruses such as lettuce necrotic yellow, sowthistle yellow vein and sonchus yellow net are membrane-bound rhabdoviruses and replicate not only in their plant host but also in their insect vector. They are therefore *circulative* and also *propagative*.

Electron microscopy of *Hyperomyzus lactucae* has shown particles of lettuce necrotic yellow virus to be located in the cytoplasm of cells of muscle, brain, fat body, mycetome, trachae, epidermis and alimentary canal (O'Loughlin and Chamber, 1967). Viruses with and without membranes were observed, indicating replication of the virus in the vector.

Sylvester and Richardson (1970) made similar observations of sowthistle yellow vein virus in *H. lactucae*. Additional evidence for virus multiplication in aphid vectors comes from serially transferring virus containing haemolymph ('blood') to non-viruliferous aphids by injection using fine glass needles. Sylvester and Richardson (1969) injected 10^5 sowthistle yellow vein virus particles into aphids. Haemolymph was then transferred in 70-fold dilutions so that, unless the virus multiplied, less than one virus particle would have been transferred by the fourth passage. However, the virus was equally efficiently transmitted at the first and sixth passages. This virus will also multiply in cell cultures of *H. lactucae* (Peters and Black, 1970) and is transmitted through the aphid eggs (Sylvester, 1969) in a process called 'transovarial transmission'.

Table 3.3   Types of aphid-borne plant viruses.

Feature	Type of virus		
	Non-persistent	Semi-persistent	Persistent
Mechanically transmissible	+	±	−
Common symptom	mosaic	yellowing	leaf roll and yellowing
Tissue of virus acquisition	epidermis	mesophyll/phloem	mesophyll/phloem
Fasting effects on transmission	+	−	−
Acquisition time	sec–mins	mins–hours	hours–days
Latent period	0	0	12 hours +
Retention of virus through moult	−	−	+
Examples	CMV	Beet yellows	Citrus tristeza
	PVY	Parsnip yellow fleck	Strawberry crinkle
	Alfalfa mosaic	Heracleum latent	Pea enation mosaic

Semi-persistent viruses. Some viruses, such as beet yellows and parsnip yellow fleck, are described as semi-persistent. They share characteristics of both non-persistent and persistent viruses (Table 3.3). Their transmission efficiency increases with increased acquisition and inoculation feeding times. There is no latent period—transmission occurs for up to 3–4 days. Pre-acquisition starving has little or no effect on transmission and virus cannot be found in the haemolymph or progeny of the vector. Virus is lost at ecdysis.

The retention of virus for 3–4 days suggests that the virus accumulates in the aphid at some retention site. Murant (1978), in studies of transmission of the anthriscus yellow virus (AYV)/parsnip yellow fleck virus (PYFV) complex by *Cavariella aegopodii*, showed virus-like particles (VLP) embedded in a matrix of material associated with the ventral lining of the posterior part of the pharynx. The suggestion is that VLP retained in the pharynx is gradually released during feeding, and that this material moves out of the aphid into the plant because of backflow from the rather inefficient oesophageal valve. Evidence of such backflow is given by Garrett (1973) and Harris and Batt (1973). This mechanism would explain transmission by non-persistent as well as semi-persistent viruses.

Specificity of transmission. One of the characteristics that distinguishes vector transmission from that which is merely mechanical is specificity. TMV for example, a highly infectious virus, is taken up by aphids but is not transmitted. *Myzus ornatus* can 'select' cauliflower mosaic virus from a mixture of cauliflower mosaic and cabbage black ringspot; on the other hand, *M. persicae* cannot distinguish the two. Similarly, this aphid transmits both henbane mosaic and turnip mosaic from mixed infections, whereas the aphid *Brevicoryne brassicae* transmits only turnip mosaic.

Barley yellow dwarf virus (BYDV) can be transmitted by about 14 species of aphid and is a persistent circulative virus. Isolates from different parts of the country are transmitted by different aphids. In the UK, an isolate from Hertfordshire is transmitted by *Rhopalosiphum padi* but not by *Metapolophium dirhodum*. This latter aphid can transmit the Bedfordshire strain but *R. padi* cannot. In the USA, *Toxoptera graminum* from different localities transmit different isolates of BYDV. The reason for specificity is not understood. In some cases virus may not be acquired from particular hosts because aphids do not probe correctly to acquire virus. When virus is acquired it may not be retained by the vector for long enough to permit transmission, or it may be inactivated, or retained too tenaciously to allow transmission. Kassanis and Govier (1971) showed that with the non-persistent virus PVY, pure virus was not transmitted unless insects had initially fed on infected plants—the virus in such plants having been inactivated by UV light. They suggest that some 'transmission factor' is picked up from the plants. Aphids allowed to acquire the transmission factor from these UV treated plants could subsequently transmit potato virus C and potato aucuba mosaic virus, neither of which is normally aphid-transmitted.

Helper viruses. A remarkable phenomenon in which one virus depends on another for its transmission by aphids was demonstrated by Smith (1946). Tobacco rosette disease is caused by infection of tobacco by two distinct viruses, tobacco mottle and tobacco vein distorting viruses. Tobacco mottle is sap-transmissible and not transmitted by *M. persicae.* Tobacco vein distorting virus is not sap-transmissible but is transmitted in persistent manner by *M. persicae.* In the presence of tobacco vein distorting virus, the mottle virus is capable of being acquired and transmitted by *M. persicae.* Tobacco mottle virus is therefore said to require a 'helper' virus. Table 3.4 lists some other viruses that need helpers.

There seem to be different reasons for dependence of viruses on helpers. Consider the parsnip yellow fleck (PYFV) and anthriscus yellow (AYV) situation where PYFV is sap-transmitted and AYV is not. In a mixture, PYFV can be isolated by mechanical transmission, and AYV by using *Myzus persicae.* PYFV can become aphid-transmissible if aphids first acquire AYV and then acquire PYFV. If, however, PYFV is acquired first,

Table 3.4 Viruses and their 'helpers'.

Virus	Helper virus	Type of virus*	Aphid vector
Tobacco mottle (ToMot)	Tobacco vein distorting (TVD)	P	*Myzus persicae*
Carrot mottle (CMoV)	Carrot red leaf (CRL)	P	*Cavariella aegopodii*
Parsnip yellow fleck (PYF)	Anthriscus yellow (AY)	SP	*C. aegopodii*
Potato aucuba mosaic (PAM)	PVY, PVA or related virus	NP	*M. persicae*
Bean yellow vein-banding (BYVB)	Pea enation mosaic (PEMV) Bean leaf roll	P	*M. persicae*

* P = persistent.
  NP = non-persistent.
  SP = semi-persistent.

no transmission of PYFV occurs, only that of AYV. These observations suggest that, to assist PYFV to be aphid-transmitted, AYV must induce some helper substance which alters PYFV. The presence of AYV alone is not sufficient.

The relationship between carrot mottle (CMoV) and its helper, carrot red leaf virus (CRLV), however, is different. Aphids are viruliferous and transmit CMoV only from plants with mixed infections. CMoV is not transmitted when aphids are allowed to feed sequentially on plants singly infected with CRLV and CMoV. Nor is the virus transmitted when aphids already carrying CRLV feed on, or are injected with, partially purified CMoV. Aphids injected with haemolymph from viruliferous aphids are able to transmit CMoV. Observations of the CMoV virions indicate the reasons for these curious facts. CMoV particles produced in singly infected plants have ether-sensitive outer lipid envelopes. CMoV from viruliferous aphids and from CRLV-infected plants are not ether-sensitive. Thus CMoV infective particles in mixed infections from aphids and from plants differ from CMoV particles from single infections. It has been suggested that particles from mixed infections consist of CMoV nucleic acid coated in CRLV protein, such particles being resistant to degradation by aphids. CMoV from single infection is coated with CMoV protein and is sensitive to aphid enzymes, being destroyed and not transmitted.

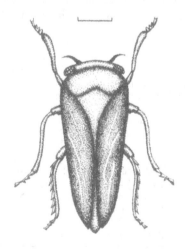

Figure 3.9　Leaf hopper of type involved in plant virus transmission (scale = 1 mm).

3.4.2 Leaf hoppers and plant hoppers

These insects (Fig. 3.9) are second only to aphids in importance as vectors of plant viruses. They frequent herbs and shrubs and move when disturbed by hopping or by flying. They have sucking mouthparts like aphids but pierce plants much more rapidly, causing more damage to plant tissues. Viruses are transmitted by leafhoppers in the semi-persistent and persistent manner.

Rice tungro and maize chlorotic dwarf (MCD) viruses are semi-persistent in their vectors. Transmission of MCD occurs after as little as 2 hours' acquisition access time and 2 hours' inoculation access time with the hopper *Graminella nigrifrons*. The virus is transmitted for several days by both nymphs and adults but is not passed transovarially.

Some persistent viruses are simply circulative in the hopper vector. Beet curly top virus, for example, may be retained by *Circulifer tenella* for long periods but does not multiply in the vector, nor is it transmitted through the egg. Wound tumour virus (WTV) however, is transmitted by *Agallia constricta* in the propagative manner. The virus infects the intestinal tract, haemolymph and salivary glands in sequence over a period of two weeks at 27°C. Peak concentrations of 10^9 virions per leafhopper are reached in four weeks. The virus is transovarially transmitted. Proof of replication was established by injecting hoppers with WTV and maintaining them on alfalfa plants that do not support virus replication. After 2–4 weeks, extracts from these hoppers were transferred by injection to other hoppers. This serial transfer was repeated and on the seventh transfer the hopper could still infect plants. Since the dilution would have been 1 in 10^{18}, multiplication of the virus in the hopper must have occurred.

Another virus, rice dwarf (RD), is transmitted more efficiently by nymphs of the leafhopper *Nephotellix cincticeps* than by adults. The incubation period for RD is 12–25 days, then transmission is for the life of the hopper. Virus passes through the egg but the resulting nymphs die prematurely.

3.4.3 Whiteflies

Whitefly-borne virus diseases, particularly of legumes, are of considerable importance in the tropics. These insects (Fig. 3.10) multiply to great numbers on the underside of leaves, as most glasshouse owners know! The whiteflies fly from plant to plant when disturbed. Viruses are acquired from the phloem of plants during feeding; they then circulate in the insect but do not appear to multiply. Successful acquisition access times may be as low

Figure 3.10 Whiteflies—vectors of plant viruses. Top, whitefly on cucumber; inset, single whitefly. Bottom, whitefly on tomato. Photographs kindly supplied by ICI Ltd, Plant Protection Division.

Table 3.5 Some whitefly-transmitted viruses.

Virus	Shape	Size (nm)	Vector
Abutilon mosaic	Geminate	20 × 30	*Bemisia tabaci*
Bean golden mosaic	Geminate	20 × 30	*B. tabaci*
Sweet potato mild mottle	Elongated	950	*B. tabaci*
Tobacco leaf curl	Geminate	20 × 30	*B. tabaci*
			Aleurotrachelus socialis

as 6 minutes, but transmission efficiency increases with increased acquisition feeding. The latent periods last from 4 to 48 hours and transmission continues for up to 20 days, but viruses are not transovarially transmitted.

Table 3.5 lists some whitefly-transmitted viruses, many of which are geminate, i.e. paired DNA virus particles, although sweet potato mild mottle virus is filamentous. Insects may acquire and transmit two viruses simultaneously, and a number of workers have shown that female whiteflies are more efficient vectors than males.

Figure 3.11 *Planococcus citri*—the citrus mealy bug vector of plant viruses. Three insects are shown, one producing a droplet of honeydew (scale = 1 mm).

3.4.4   Mealy bugs (Fig. 3.11)

These are responsible for the transmission of cacao swollen shoot virus, cacao mottle and cacao trinidad virus. Infectivity of mealy bugs increases with increased acquisition feeding. These viruses are circulative and there is no passage through the insect eggs. The viruses are acquired by sucking, stylet-like mouthparts. Nymph and adult species of *Planococcus* are equally efficient vectors of cacao swollen shoot virus, but male insects are unable to transmit. Nymph and juvenile insects are most active and spread virus within crops, longer distance transmission coming about by wind. Interestingly, infective mealy bugs may also be carried about by ants, although this happens only when an ant-attended colony is disturbed.

3.4.5   Thrips

Transmission of tomato spotted wilt (TSWM) and tobacco ringspot (TRV) is reported to be by species of thrips. TRV is transferred only by nymphs and not by adults. TSWV, on the other hand, is acquired by the larvae of *Thrip tabaci* but not by adults, whereas only adults transmit. The shortest acquisition time is 15 minutes; there is a latent period of 4–10 days; adults are maximally effective after 22–30 days and retain virus for life. Virus is not transmitted to progeny.

3.4.6   Beetles

Beetles (Coleoptera) are chewing insects (Fig. 3.12) unlike the Hemiptera (Homoptera). Some beetle-transmitted viruses are listed in Table 3.6. Two

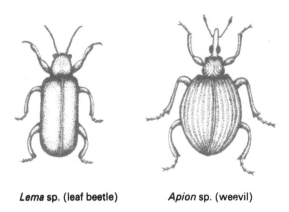

Lema sp. (leaf beetle)   *Apion* sp. (weevil)

Figure 3.12   Coleoptera involved in plant virus transmission.

Table 3.6 Some beetle-transmitted viruses.

Virus	Vector
Cowpea mosaic	Ceratoma trifurcata
Squash mosaic	Acalymna trivittata
	Diabrotica undecimpunctata
Radish mosaic	D. undecimpunctata
	Phyllotreta cruciferae
	Epitrix hirtipennis
Echtes ackerbohnenmosaik	Apion vorax
	Sitona lineatus (weevils)
Turnip yellow mosaic	Phyllotreta sp.
	Phaedon cochleariae
Wild cucumber mosaic	A. trivittata
Southern bean mosaic	C. trifurcata
Cocksfoot mottle	Lema melanopa
Turnip crinkle	Phyllotreta spp.
	Psylliodes spp.

main super-families of Coleoptera transmit plant virus. These are the Chrysomeloidea (family Chrysomelidae), the leaf beetles, and the Curculionoidea (family Curculionidae), the weevils. Transmission is more than a mechanical event, since the virus persists in the insects and can be detected in the haemolymph of chrysomelid beetles. In many instances however, it is thought that virus is inoculated following regurgitation of sap from the fore-gut. Viruses such as broad bean stain and echtes ackerbohnenmosaik cannot be detected in weevil haemolymph, and weevils have not been observed to regurgitate food (Cockbain et al., 1975). Retention of virus for up to 8 days suggests that transmission is more than a simple mechanical non-persistent event.

3.4.6 Mites

Mites (Acarina) are a group related to the spiders (arachnids). Plant mites feed by sucking the contents from plant cells. Two families, the Tetranychidae and the Eriophyidae transmit viruses; some mite-transmitted viruses are listed in Table 3.7.

The Tetranychidae—spidermites—are exclusively plant parasites and important pests of agricultural crops. These mites are 8-legged, up to 0.88 mm long, oval or pear-shaped. Two species, Tetranychus urticae (Koch) and T. telarius (L.), have been reported to transmit PVY in the non-persistent manner.

Table 3.7   Some plant viruses transmitted by mites.

Virus	Vector
Wheat streak mosaic	*Aceria tulipae*
Rye grass mosaic	*Abacarus hystrix*
Agropogon mosaic	*A. hystrix*
Prunus latent	*Vasates fockeni*
PVY	*Tetranychus urticae* ?
PVY	*T. telarius* ?
Tobacco ringspot	*Tetranychus* spp.

Eriophyids are the more important vectors of plant viruses. These mites are about 0.2 mm long, 4-legged and elongated (Fig. 3.13). Most are host-specific, although some, such as *Aceria tulipae*, colonize more than one plant family but are limited to certain species within those families. Although possessing two pairs of legs, their main method of spread is by wind currents. They swarm on leaves, then leap into the wind to be carried away. Eriophyids have two nymphal instars followed by a 'pseudopupal' stage. Development from egg to adults takes 6–14 days.

Nymphal stages only of *A. tulipae* can acquire wheat streak mosaic virus. Acquisition and transmission take about 15 minutes each, and virus is maintained through the moult. EM studies show virus to be confined to the gut, and there is no evidence of replication in the vector even though the virus is retained for nine days in mites kept on virus-immune plants. There is no evidence of virus passing to progeny.

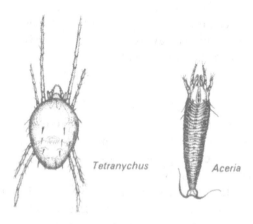

Figure 3.13   Two types of mite vectors of plant viruses. Left, *Tetranychus urticae*, vector of PVY; right, *Aceria tulipae*, vector of wheat streak mosaic virus.

3.5 Transmission by soil vectors

A number of viral diseases of plants were found to arise even when plants were kept free of possible insect vectors. In these cases the soil was suspected as the source of virus, and a number of diseases were described as resulting from *soil-borne viruses*. Cadman (1963) divided such virus diseases into two groups—those losing infectivity when soil was allowed to dry at 20°C for a week or more, and those where soil remained infective on drying for several weeks or years. Cadman realized that these differences reflected properties of the vector and not of the viruses. Those viruses in the first group have subsequently been found to be transmitted by nematodes and the second group by fungi.

3.5.1 *Nematodes or eel worms* (Fig. 3.14)

These feed on plant roots by inserting stylets in a manner similar to sucking insects, and virus is actively introduced into cells by the worms.

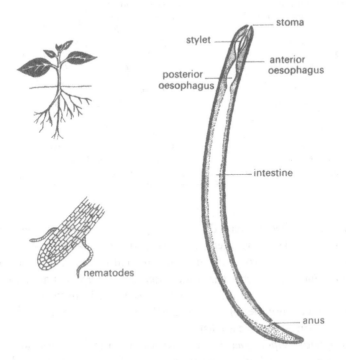

Figure 3.14 Nematode vectors of plant viruses. Left, nematodes attached to plant root; right, diagram of a nematode.

Table 3.8 Some nematode-transmitted viruses.

Viruses	Nematodes
Nepoviruses	
Arabis mosaic (typic strain)	*Xiphenema diversicaudatum,* *X. coxi*
Arabis mosaic (hop strain)	*X. diversicaudatum*
Grape fan leaf	*X. index, X. italiae*
Cherry leaf roll	*X. coxi, X. diversicaudatum,* *X. vuiltenezi*
Tobacco ringspot	*X. americanum, X. coxi*
Tomato ringspot	*X. americanum*
Tomato black ring	
(English and German strains)	*Longidorus attenuatus*
(Scottish strain)	*L. elongatus*
Raspberry ringspot	
(English strain)	*L. macrosoma, L. elongatus*
(Scottish strain)	*L. elongatus*
Tobraviruses	
Pea early browning	*Paratrichodorus anemone*
(English isolates)	*Trichodorus primitivus,* *T. viruliferus*
(Dutch isolates)	*P. pachydermus, P. teres*
Tobacco rattle	*P. anemone, P. nanus, P. pachydermus,*
(European isolates)	*P. teres, T. cylindricus, T. minor,* *T. primitivus, T. similis,* *T. viruliferus*
(American isolates)	*P. allius, P. christiei, P. porosus*
Other viruses	
Brome mosaic	*X. diversicaudatum, L. macrosoma*
Carnation ringspot	*X. diversicaudatum, L. elongatus*
Prunus necrotic ringspot	*L. macrosoma*

Hewitt *et al.* (1958) were the first to positively show that eelworms transmit virus. These workers demonstrated that 'fan leaf disease' of grapevine was spread from infected plants to healthy plants growing in the same container only if the nematode *Xiphenema index* was added to the soil. No disease occurred if *X. index* was added to healthy plants alone. Since that time a number of viruses have been shown to be transmitted by nematodes, some of which are listed in Table 3.8.

Four genera, *Xiphenema, Longidorus* (Longidoridae), *Trichodorus* and *Paratrichodorus* (Trichodoridae), transmit plant viruses. Longidoridae are relatively large nematodes, adults being 3 mm or more long, and feed by inserting stylets below the root epidermal layers. Trichodoridae are only 1 mm long and feed on epidermal cells, transmitting straight tubular viruses of the tobacco rattle (tobravirus) type. *Xiphenema* and *Longidorus* transmit isometric viruses of the nepo group—a name derived from *nema*tode transmitted *p*olyhedral viruses.

Arabis mosaic and tomato ringspot can be acquired after one hour's feeding on infected plants, and transmitted in a similar period. *X. divesicandatum* retains arabis mosaic for at least 31 days when kept in fallow soil, and for at least eight months when kept on virus-immune varieties of raspberry (Harrison and Winslow, 1961). Larvae and adult nematodes can transmit but the virus is not passed to progeny, neither is the virus retained after moulting.

Specificity of transmission occurs; for example, the Scottish strain of raspberry ringspot is transmitted only by *Longidorus elongatus*, and the serologically distinct English strain by *L. macrosoma* and *L. elongatus*. Other examples are given in Table 3.8.

Nematodes are not very mobile but tend to intensify virus within crops, viruses have entered crops from weeds where transmission is by seed.

3.5.2. Fungi

Viruses that persist in soil for long periods were found to be transmitted from one plant to another by fungi. Two orders of fungi, the Chytridiales and the Plasmodiophorales, acquire virus from virus-infected plants.

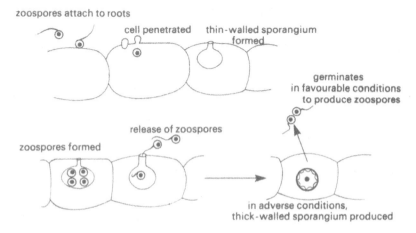

Figure 3.15 Diagrammatic life cycle of *Olpidium brassicae*, vector of tobacco necrosis virus.

Table 3.9 Some plant viruses transmitted by fungi.

Virus*	Vector	Longevity in spores (yrs)
Tobacco necrosis (I)	Olpidium brassicae	—
Soil-borne wheat mosaic (I)	Polymyxa graminis	3
Cucumber necrosis (I)	O. cucurbitacearum	—
Potato mop top (E)	Spongospora subterranea	2
Beet necrotic yellows (E)	P. betae	4
Wheat spindle streak (FE)	P. graminis	8
PVX (FE)	Synchytrium endobioticum	—

* Particle shape: I = isometric; E = rigid, elongated; FE = flexuous, elongated.

When these soil-inhabiting fungi produce motile stages, they contain virus and can then infect further plants (Fig. 3.15). The fungi, in adverse conditions, produce highly resistant virus-containing resting spores and this explains the persistence in soils for months or years of some of the viruses. Table 3.9 shows the length of survival of viruses in spores, although some viruses, such as TNV and cucumber necrosis, may be present only on the surface of resting spores.

Fungi show some specificity of transmission, since *Olpidium brassicae* will transmit TNV but not cucumber necrosis virus, whilst *O. cucurbitacearum* transmits cucumber necrosis but not TNV.

CHAPTER FOUR

PLANT VIRUS STRUCTURE

In this chapter the biochemical structure of plant viruses will be described, followed by an examination of virus morphology and architecture.

4.1 Biochemistry of plant viruses

Viruses consist of nucleic acid coated with protein; some may also contain lipid and other compounds. Fig. 4.1 shows something of the classification of plant viruses. The major subdivision is based on the type of nucleic acid content of the viruses.

4.1.1 *Viral nucleic acids*

These consist of chains of nucleotides (Fig. 4.2) linked by phosphodiester links between the 5′ hydroxyl of the sugar of one nucleotide and the 3′ hydroxyl of another. The chain will therefore have a 5′ hydroxyl free at one end (this is known as the 5′ end), and a 3′ hydroxyl free at the other, giving a 3′ end. In some cases the ends are joined, giving circular nucleic acid molecules.

The chemical bases involved in RNA and DNA are identical, except that thymine of DNA is replaced by uracil in RNA (Table 4.1). Furthermore, the deoxyribose sugar of DNA is replaced by ribose in RNA.

RNA and DNA can each exist either as single strands or as double-stranded structures. In general, DNA is more usually double- and RNA single-stranded. Fig. 4.2 shows the structure of part of the double helix formed from two strands of DNA. This structure is produced by pairing between purine and pyrimidine bases, adenine (A) pairing with thymine (T), and cytosine (C) pairing with guanidine (G). Thus the nucleotide

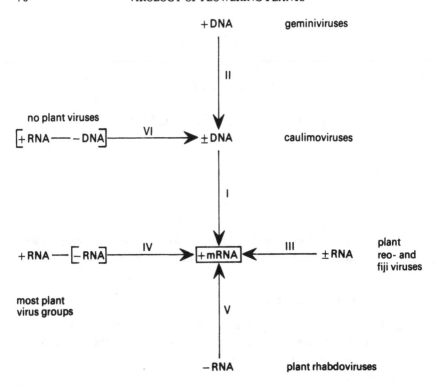

Figure 4.1 Composite diagram of pathways of mRNA synthesis used by plant viruses. Based on the scheme by Baltimore (1971). mRNA defined as +RNA.

sequence in one chain is complementary to that in the other. In double-stranded RNA, C and G pair, but A is paired with uracil (U) since there is no thymine.

The sequence of bases in the nucleic acid of viruses is used as a code to determine the amino acid sequence, and therefore the characteristics, of the virus proteins. The properties and behaviour of the virus will depend to some extent on these proteins. Three nucleotides code for each amino acid, and other nucleotide sequences terminate production of proteins and polypeptides (Table 4.2). The nucleic acid of the virus is known as the virus *genome*.

Viruses may contain their nucleic acid as a single continuous strand, or the nucleic acid may be present as two or more pieces. Such viruses are said to have a *divided* or *multipartite* genome. The different nucleic acids may be present in separate particles that can be distinguished by size or density.

Viral RNAs show some distinct structural features, which are briefly described below.

3′ terminal end. The 3′ end of eukaryotic and some prokaryotic messenger RNA (mRNA – see section 5.1.2) is terminated by a sequence of many

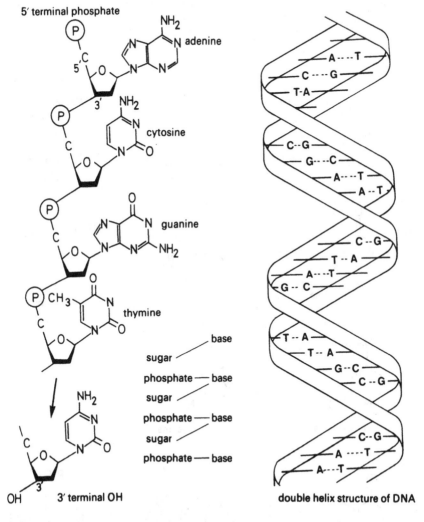

Figure 4.2 Structure and arrangement of the components of nucleic acids. The structure shown is that of deoxyribose nucleic acid (DNA). In ribose nucleic acid (RNA), uracil replaces the base thymine and the sugar is ribose. Details of chemical components are given in Table 4.1.

Table 4.1 Chemical components of nucleic acids.

	DNA only	DNA and RNA	RNA only
Nitrogen bases			
Purines		Adenine, guanine	
Pyrimidines	Thymine	Cytosine	Uracil
Pentose sugar	Deoxyribose		Ribose
Phosphate		H_3PO_4	

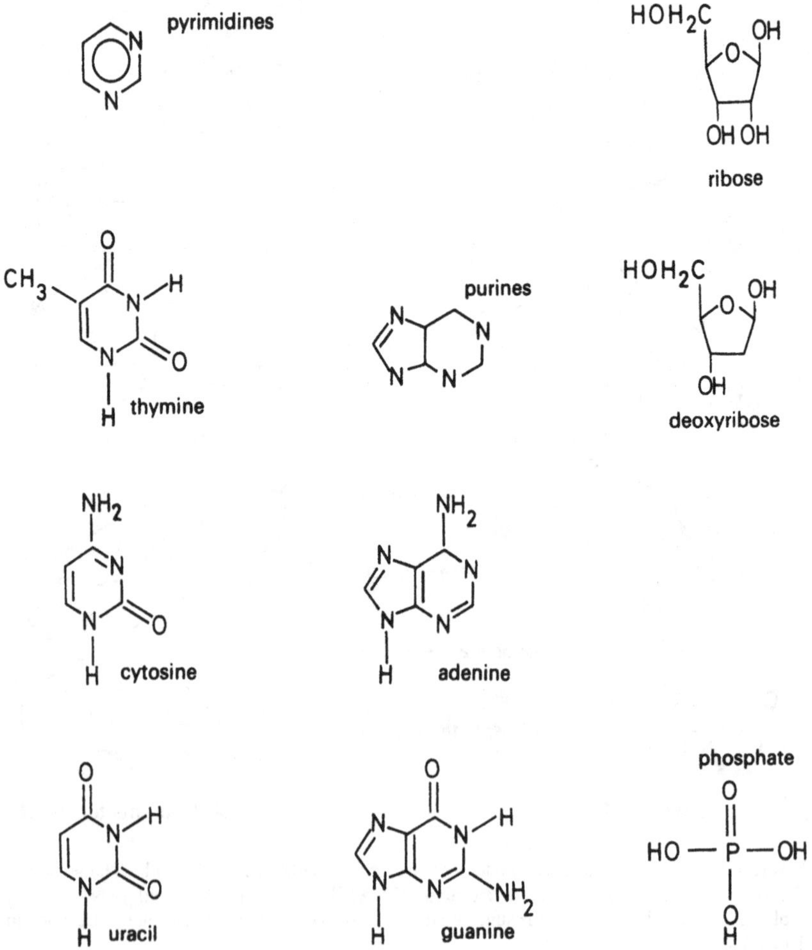

pyrimidines

ribose

CH_3 — thymine

purines

deoxyribose

cytosine

adenine

uracil

guanine

phosphate

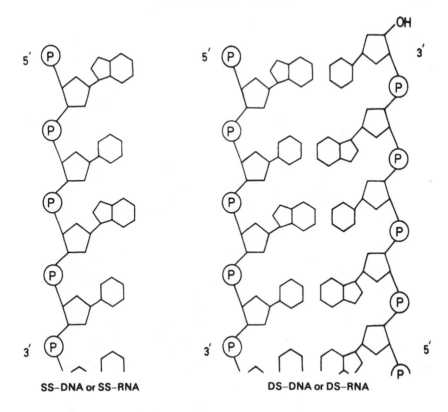

5′ · 3′

5′ · 3′ · 5′

OH

SS–DNA or SS–RNA

DS–DNA or DS–RNA

Figure 4.3 Single-strand (SS) and double-strand (DS) nucleic acids.

Table 4.2 The base codes, or codons, for amino acid incorporation into protein.

Amino acid	Codons	Amino acid	Codons
Alanine (Ala)	GCX	Leucine (Leu)	UUA, UUG, CUX
Arginine (Arg)	CGX, AGA, AGC	Lysine (Lys)	AAA
Asparagine (Asn)	AAU, AAC	Methionine* (Met)	AUG
Aspartic acid (Asp)	GAU, GAC	Phenylalanine (Phe)	UUU, UUC
Cysteine (Cys)	USU, UGC	Proline (Pro)	CCX
Glutamic acid (Glu)	GAA, GAG	Serine (Ser)	UCX, AGU, AGC
Glutamine (Gln)	CAA, CAG	Threonine (Thr)	ACX
Glycine (Gly)	GGX	Tryptophan (Trp)	UGG
Histidine (His)	CAU	Tyrosine (Tyr)	UAU, UAC
Isoleucine (Ile)	AUU, AUC, AUA	Valine* (Val)	GUX (GUG*)

* These also act as initiation codons.
Termination codons UAA, UAG and UGA.
X = stands for any one of four bases.

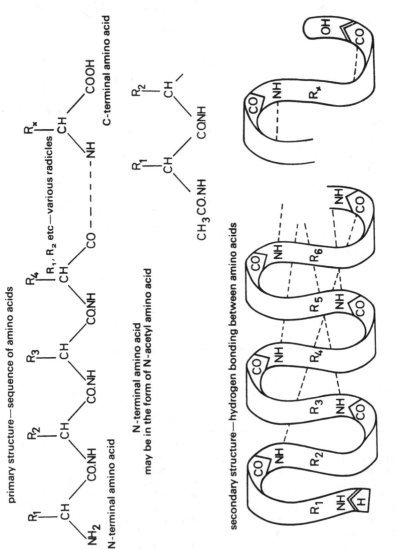

Figure 4.4   Protein structure.

adenylic acid molecules. The function of these polyadenylic acid (polyA) sequences is not known. Such polyA is found in the RNA of animal viruses but less commonly in plant viruses. It occurs in como-, nepo-, hordei- and rhabdoviruses. 3′ terminal polyA is also found in the mRNA associated with infection by cauliflower mosaic, a DNA virus.

5′ capping. The 5′ end of many, but not all, mRNAs is terminated by a methylated guanosine molecule. This 'cap' is thought to act as an aid in the translation of the mRNA by influencing its binding to ribosomes or initiation factors. Similar caps are found in viral RNA of a number of plant viruses including tobaviruses, tymoviruses, bromoviruses, rhabdoviruses and alfalfa MV.

tRNA-like structure. Part of the viral RNA of toba-, tymo, cucumo- and bromoviruses can bind specific amino acids at the 3′ end on treatment with aminoacyl-tRNA synthetase; in this respect it resembles transfer RNA (tRNA—see section 5.1.2). Although part of the viral RNA may be looped, it does not closely resemble the clover-leaf configuration of tRNA. Eggplant mosaic virus contains both genomic RNA and smaller molecular weight particles with tRNA properties (Bouley *et al.,* 1976).

4.1.2 Viral proteins

Like other proteins, these are composed of amino acids linked by peptide bonds. The sequence of amino acids constitutes the *primary structure* of the protein. The amino acids may produce a helical structure (Fig. 4.4) and this is called the *secondary structure* of the protein. A further level of organization comes about when the protein folds to give *tertiary structure.*

As can be seen in Fig. 4.4, the amino acids which terminate the sequence may have a free amino group (N-terminal group), or a C-terminal group in the form of a carboxyl group. The complete amino acid sequences for proteins from a number of viruses are now known. A particular point of interest is that the N-terminal amino acid is often an *N*-acetyl amino acid (Fig. 4.4), the reason for which is not understood; it may be that it protects the viral protein from attack by host cell proteinases.

In cauliflower mosaic virus, some of the protein is phosphorylated at the amino acids serine and threonine (Hahn and Shepherd, 1980).

Table 4.3 gives the molecular weight (particle weight) of some viruses, their nucleic acid (NA) and their protein. Molecular weights (MW) are given throughout the text in daltons. If even the smallest viruses were

Table 4.3 Molecular weights of some virus particles, their nucleic acid and protein sub-units.

Virus	Shape†	Particle MW × 10⁶	Nucleic acid MW × 10⁶	Protein × 10⁴
Tobacco mosaic	E	39.4	RNA 2.05	1.7
Potato virus X	FE	35.0	RNA 2.10	2.2 or 3.0*
Cucumber mosaic	I	5.8–6.7	RNA 1.0	3.2
Tobacco necrosis	I	7.0	RNA 1.3–1.6	2.26–3.35**
TNV satellite	I	1.97	RNA 0.28	2.3
Turnip yellow mosaic	I	‡3.6(T) 5.4(B)	RNA 2.0	2.0
Cauliflower mosaic	I	2.3	DNA 5.0	3.2–7.0***
Bean golden mosaic	G	2.6	DNA 0.75	3.1

† E = elongated; FE = flexible elongated; I = isometric; G = geminate (see 4.6.2).
‡ T = top; B = bottom components.
* Value dependent on treatment.
** Various reported values.
*** Range of protein sizes.

composed of a single protein, such a compound would be extremely large. For example, tobacco necrosis, with an MW of 6.3×10^6, is composed of 81 % protein and 19 % nucleic acid, so a single protein would have an MW of about 5.3×10^6; similarly, the protein of TMV would need to have an MW of about 38×10^6. These molecules are too large to be coded for by the nucleic acid, and it is now known that the protein component of all viruses is composed of a number of identical subunits. The molecular weights of subunits for different viruses vary from 15 000–60 000.

4.1.3   Other chemical components of viruses

Some viruses contain substances other than protein and nucleic acid. Matthews (1970) draws attention to the fact that viruses are hydrated, and crystals of tomato bushy stunt and turnip yellow mosaic virus contain 47 % and 58 % respectively of water.

Carbohydrates are associated with the protein of sonchus yellow net and other rhabdoviruses, and are present as glycoprotein in the virus envelope. Tomato spotted wilt, a membrane-bound virus with icosahedral symmetry, is said to contain 7 % carbohydrate (Best, 1968). The coat proteins of barley stripe mosaic and European wheat streak mosaic viruses are also glycoproteins (Partridge *et al.*, 1974). Lipid is also found in membrane-bound viruses and constitutes 15–20 % of the virus particle.

Polyamine, such as spermidine, makes up about 1 % by weight of turnip yellow mosaic virus. Spermidine (Fig. 4.5) like other polyamines, may

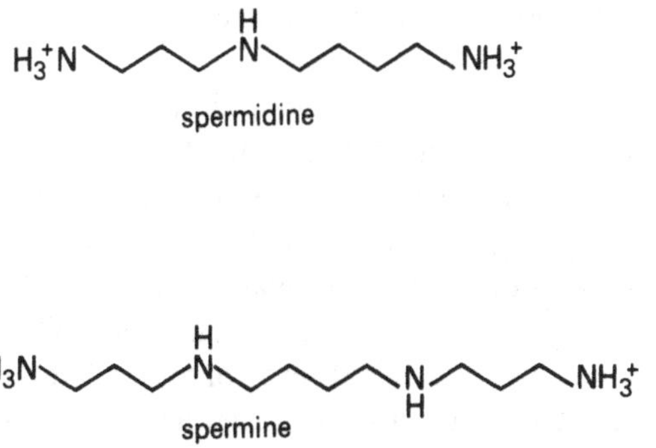

spermidine

spermine

Figure 4.5 Two examples of polyamines found in plant viruses.

interact with and neutralize phosphate groups on nucleic acids, or may stabilize folded nucleic acid molecules.

Metallic ions, such as calcium, sodium and magnesium, have been found in purified virus preparations. Such metals may be bound either to protein or nucleic acid or both. The function of such ions is not understood, but they may have stabilizing effects on the virus particles (Cohen and McCormick, 1979).

Enzymes are less commonly found in plant viruses than in those from animals. RNA polymerases have been recorded from wound tumour virus, and the rhabdoviruses lettuce necrotic yellow, broccoli necrotic yellow, and sonchus yellow net virus (Jackson and Christie, 1979).

4.2 Virus architecture

Viruses carry their genetic information protected by a protein coat or *capsid*. The amount of genetic information supplied by the RNA or DNA is limited, and this restricts the complexity that viruses can attain. As we have seen, the protein for many viruses is confined to one type which is replicated in the structure many times. The advantage of such construction leads to further economy of genetic information by facilitating *self-assembly*. Virus proteins, if separated from their nucleic acid, will under suitable conditions assemble as virus-like particles or *empty particles*. Proteins from rod-shaped viruses assemble to form rods and those from icosahedral particles assemble to produce icosahedral particles. Self-assembly is a feature of the individual protein subunits and is determined by the shape and structure of the subunit protein; in this way there is economy of genetic information required for building viruses. Any incorrectly-formed protein will be rejected during the virus assembly process.

The linking of protein subunits and nucleic acid comes about in order to produce a structure of lower energy content than the free components. Such a package of components will collect together to give a *minimal energy* structure with *maximum stability*. This stability can best be achieved if individual protein subunits are in similar 'environments', that is, bonding between individual proteins is identical or *equivalent*. Where molecules (or atoms) are in exactly equivalent environments, we have crystalline structures. Viruses are not crystalline, however; they contain one or just a few nucleic acid components and rather more protein units. The proteins may be considered as surface crystals rather than three-dimensional periodic crystals (Casper, 1964). The condensation of these protein units around the nucleic acid means that viral proteins may not be in exactly equivalent environments. Casper and Klug (1962) describe the situation by the term *quasi-equivalence*, where bonds between sub-units show small non-random variations from regular bonding patterns, leading to more stable structures than strictly equivalent bonding. The most probable minimum energy designs for surface crystals constructed of a number of units are tubes with helical or cylindrical symmetry, and closed shells with icosahedral symmetry (see 4.2.2). Quasi-equivalent bonding is a geometrical necessity of icosahedral shells, but simply a convenience for helical structure. The principles of construction are the same for both helical and icosahedral or isometric particles.

Differences between helical and isometric viruses come about as a result of differences in nucleic acid. Helical viruses have lengths of nucleic acid making intimate contact with their surface proteins. Isometric viruses have coiled helices of nucleic acid with intra-molecular bonds giving a globular structure with rather less contact between nucleic acid and proteins.

4.2.1 *Elongated viruses*

For many years it was assumed that all viruses were spherical in shape. In 1933, however, Takahashi and Rawlins (see Bawden, 1956) showed that clarified sap from TMV-infected plants showed optical properties indicating the presence of rod-shaped particles not to be found in sap from healthy plants. The rod-shaped nature of TMV was confirmed later, using the electron microscope. Many viruses have since been shown to have helical symmetry—such viruses can be either rigid tubes or long flexuous tubes, and Table 4.4 lists some examples of each of these types. The best-

Table 4.4 Groups of viruses showing helical symmetry (elongated viruses).

Group	Type virus	Size (nm)	% NA
A. *Rigid tubular viruses*			
Tobamovirus	Tobacco mosaic	300 × 18	5
Tobravirus	Tobacco rattle	190 × 26	5
Hordeivirus	Barley stripe mosaic	128 × 28	4
Carlavirus	Potato virus S	650 × 12	5
B. *Flexuous tubular viruses*			
Closteroviruses	Beet yellow mosaic	1250 × 10	5
Potyviruses	Potato virus Y	730 × 11	5
Potexviruses	Potato virus X	515 × 13	6

documented of the rigid tubular viruses is tobacco mosaic virus, and details of its structure can be seen in Fig. 4.6 and Table 4.5. The RNA is helical, leaving a hollow centre. Protein subunits equivalent to single protein molecule capsomeres follow the contour of the RNA. The RNA is in fact held in a groove in each protein molecule in such a way that three nucleotides are associated with a single protein, 49 proteins being distributed in three turns of the 23 Å-pitch helix. TMV particles are 300 nm long and 18 nm in diameter. The single strand of RNA consists of 6340 nucleotides, giving a nucleic acid of molecular weight 2.05×10^6. This represents about 5 % of the particle weight. Each of the 2130 proteins is identical and composed of 158–161 amino acids, giving subunits of

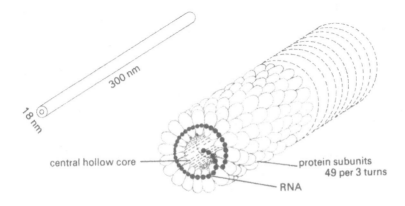

Figure 4.6 The structure of tobacco mosaic virus (a rigid rod-shaped virus).

Table 4.5 Some basic characteristics of tobacco mosaic virus.

Particle weight	39.4×10^6
RNA MW	2.05×10^6
Protein subunit MW	1.75×10^4
Number of nucleotides	6390
Number of protein	2130
Percentage protein	95
Length	300 nm
Diameter of particle	18 nm
Diameter of core	4 nm
Pitch of helix	2.3 nm
Protein subunits per turn	$16\frac{1}{3}$
Nucleotides per turn	49

molecular weight $1.7–1.8 \times 10^4$. The end of the viral tube containing the 3′ end of the RNA is convex, the other end is concave (Wilson *et al.*, 1976).

Some other tubular viruses with helical symmetry have flexuous particles. Potato virus X, for example, is 515 nm long and 13 nm wide (Fig. 4.7). The structure is very similar to TMV, there being a central hollow core and protein subunits arranged nine per turn of the helix (Richardson *et al.*, 1981). Beet yellows is a flexuous tube 1250 nm long and only 10 nm in diameter. The pitch of these helical structures is 33–37 Å (Varma *et al.*, 1968) and there is probably less intimate contact between protein subunits and RNA, resulting in a more flexible structure.

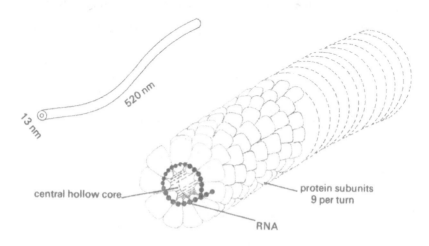

Figure 4.7 The structure of potato virus X (a flexible rod-shaped virus).

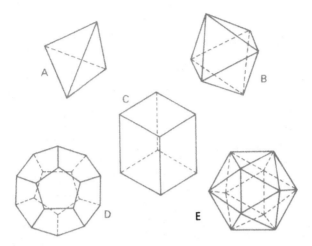

Figure 4.8 The five platonic bodies. A—tetrahedron; B—octahedron; C—cube; D—dodecahedron; E—icosahedron.

4.2.2 Isometric viruses

Electron microscope observations of many plant viruses showed them to be spherical or near-spherical in shape. When it was realized that the capsid of these viruses was composed of a number of identical protein units, the concept of spherical viruses changed.

For minimum energy requirement and maximum stability, spherical or near-spherical arrangements of subunits is ideal. Subunits arranged on the surface of a sphere with equivalence might conceivably conform to one or other of the 5 platonic bodies (Fig. 4.8). The platonic bodies can each be inscribed within a sphere in such a way that their vertexes touch the surface and these points of contact are equal distances apart. Subunits placed at these points would be *equivalent* and interact equally between themselves.

The five bodies are related by Euler's formula $(E + 2 = V + F)$ where E = number of edges, V = number of vertexes and F = number of faces. The symmetry of these bodies is rotational, and for the simplest, the tetrahedron, it is $2:3$ (Fig. 4.9); for the icosahedron it is $5:3:2$.

Crick and Watson (1956) pointed out that subunits might be arranged with centric symmetry based on a tetrahedron, an octahedron or an icosahedron. Many viruses are structured on an icosahedron with 20 faces, 12 vertexes and 30 edges.

Evidence for icosahedral symmetry in viruses came from X-ray diffraction patterns of tomato bushy stunt virus obtained by Casper (1956) and

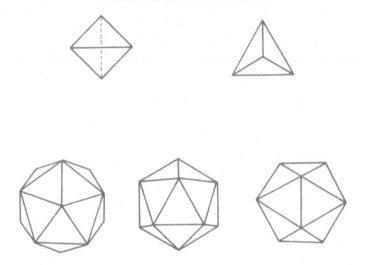

Figure 4.9 Symmetry of two of the platonic bodies. Top, tetrahedron viewed in two positions to show 2:3 symmetry. Bottom, icosahedron viewed in three positions to show 5:3:2 symmetry.

from turnip yellow mosaic virus (TYMV) (Klug *et al.*, 1957). The X-ray pattern (Fig. 4.10) showed 5:3:2-fold symmetry. Further evidence for icosahedral symmetry was found using an insect virus, tulipa iridescent virus. When this virus was prepared for electron microscopy and shadowed in two directions, the shapes of the shadows conformed to those

Figure 4.10 A copy of the precession photographic image of a tomato bushy stunt virus crystal, showing 5:3:2 symmetry of an icosahedral structure.

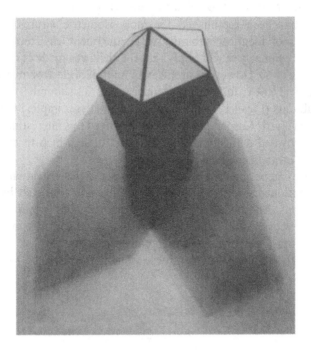

Figure 4.11 A model showing shadow shapes used to confirm icosahedral shape of viruses. Similar shadow shapes can be obtained in the electron microscope with virus particles.

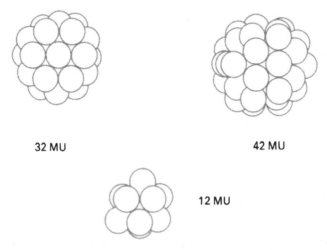

32 MU

42 MU

12 MU

Figure 4.12 Arrangement of morphological subunits (MU) in three of the simplest types of virus.

produced only by icosahedral structures (Fig. 4.11). This icosahedral symmetry has been established for most isometric viruses. Now the arrangement of identical subunits with equivalent environments with icosahedral symmetry would imply that the shell was made of 60 subunits or a multiple of 60. However, careful examination of electron micrographs showed that TYMV, for example, consisted of 32 subunits or morphological subunits (Fig. 4.12) whereas chemical analyses implied that nearer 200 were involved! Casper and Klug (1962) solved this interesting paradox and also provided a theoretical background upon which to explain the structure of other isometric viruses.

Virus size can be changed by varying the size of the individual subunits, but, as mentioned, the dimensions of subunits are restricted by genetic considerations. An alternative is to use more subunits within the framework of an icosahedral structure. Now the icosahedron is made up of equilateral triangles (Fig. 4.9) which can be increased in size to accommodate more subunits. How can these subunits be arranged to maintain

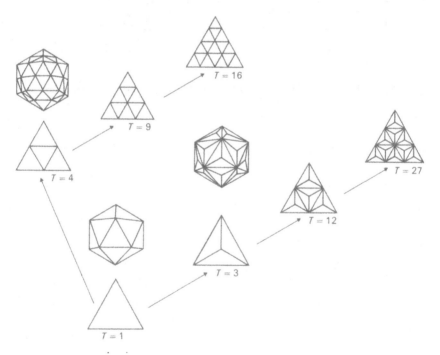

Figure 4.13 Triangulation of icosahedra to accommodate more protein subunits. Note centres of 5-fold and 6-fold symmetry. Faces triangulated for three values only in each series (see text). The three simpler icosahedra only have been drawn.

equivalent or quasi-equivalent environments? The answer can be found by subdividing the triangular faces into further triangles. Since there are 20 faces in an icosahedron, the number of sub-triangles will be $20T$ where T is the *triangulation number*. T is given by the formula $T = H^2 + HK + K^2$, where H and K are any pair of integers, or $T = Pf^2$ where P is 1, 3, 7, 13... and f is any integer. When $P = 1$ the lowest class T becomes 1, 4, 9, 16, 25 ..., and when P is 3, T is then 3, 12, 27... This is shown diagrammatically in Fig. 4.13.

If the icosahedron in Fig. 4.9 is examined it can be seen that each vertex has five-fold symmetry. Thus five proteins can be placed in equivalent environments at each vertex. Since there are 12 vertexes, 60 subunits can be accommodated on the simplest icosahedron $(T = 1)$. This feature is common for all icosahedra. The placement of further proteins will depend on how the triangular faces are sub-divided. If each face of the icosahedron is divided into four equilateral triangles $(T = 4)$, as well as the five-fold points of symmetry, there are points of six-fold symmetry. Fig. 4.13 shows that there are 30 such points. This structure has 42 centres of symmetry, 12 composed of five subunits (pentamers) each as in the simplest example, together with 30 points where groups of six subunits (hexamers) can be arranged. Such a virus will have 42 groups of subunits or capsomeres and 240 individual proteins or structure subunits. Working on the same principle, TYMV is an icosahedron with 32 morphological units composed of $12 \times 5 = 60$ subunits at vertexes, and 20×6 subunits, one on each face, giving a total of 180 structural units. These figures correspond with the

Table 4.6 Number of morphological and structural subunits and hexamer arrangement in quasi-equivalent icosahedral viruses.

f	T^*	Number of: Morph. subunits	Struct. subunits	Number of hexamers on: Faces	Edges
A. When $P = 1$					
1	1	12	60	0	0
2	4	42	240	0	1
3	9	92	540	1	2
B. When $P = 3$					
1	3	32	180	1	0
2	12	122	720	4	1
3	27	272	1620	10	2

* $(T = Pf^2$—see text).

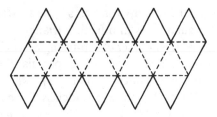

Figure 4.14 A template for constructing an icosahedron.

number of subunits calculated theoretically ($\times 60$) observed in the EM (32) and detected by chemical analysis ($\simeq 200$).

Table 4.6 shows the arrangement of morphological and structural units for different triangulation numbers. Careful consideration of the numbers of hexamers in different particles shows that they are located either on the faces ($20 \times$) or on the edges ($30 \times$) and in various combinations of these. It is instructive to make models of icosahedra (see Fig. 4.14) and by triangulating the faces to determine the number and position of the hexamers for various triangulation numbers. Details of the structure of turnip yellow mosaic virus are given here as this is one of the viruses used to establish the basic structure of isometric viruses.

TYMV has icosahedral symmetry and is 28–29 nm in diameter. Electron micrographs of particles negatively stained with uranyl acetate show 32 morphological subunits. Twelve of the morphological units, situated at vertexes, are pentamers; the remaining 20 are hexamers, giving 180 protein subunits. The protein is of one type, of MW 20 133, and contains 189 amino acid residues. The RNA is single-stranded but details of its arrangement in the protein shell are not known. The single length of nucleic acid has a MW of 2.0×10^6 containing about 6400 nucleotides and constituting 35 % of the weight of the virus.

Density gradient centrifugation of preparations of TYMV produces several bands, indicating different types of virus particles. Some remain near the top of the centrifuge tube and consist of empty protein shells (T or top components). Infective nucleoprotein and virus particles constitute B or bottom components. Non-infective particles with small sub-genomic RNA have densities intermediate between T and B components.

Other viruses with isometric symmetry are listed in Table 4.7. Many groups have more than one size of nucleic acid, although with exception of como- (Fig. 4.15), caulimo- and tombusvirus groups, each has only a single species of protein.

Table 4.7   Groups of isometric viruses.

Group	Type example	Shape*	Diam. nm	No. M.U.	Protein S.U.	No. RNA species
Bromoviruses	Brome mosaic	Ic	26	32	180	4
Comoviruses	Cowpea mosaic	Ic	24	32	120	2
Cucumoviruses	Cucumber mosaic	Ic	28	32	180	4
Nepo	Tobacco ringspot	Ic	28	12	60	2
Tombus	Tomato bushy stunt	Ic	30	90	180	1
Tymo	Turnip yellow mosaic	Ic	28	32	180	1
Ilar	Prunus necrotic ringspot	Is	28–30	?	—	1
Luteo	Barley yellow dwarf	Is(O)	20–24	?	—	(4)
Phyto/reo	Wound tumour	IsD	70	?	—	12
Fiji	Sugarcane fiji disease	IsD	70	?	—	10
Southern bean mosaic		Is	28	?	?	1
Tobacco necrosis		Ic	26	12	60	1
Satellite		Ic	17	12	60	1
Maize chlorotic dwarf		Is	30	—	—	1
Caulimo	Cauliflower mosaic	IsD	50	—	450+	1(DNA)

* Ic = icosahedral
  Is = isometric
  IsD = isometric, double shelled
  Is(O) = isometric, octahedral
  M.U. = morphological units
  S.U. = structural units.

Figure 4.15   The structure of turnip yellow mosaic virus showing protein subunits (180) in pentamers (5's) and in hexamers (6's) making up the 32 morphological subunits.

Viruses such as prunus necrotic ringspot, barley yellow dwarf, southern bean mosaic and maize chlorotic dwarf, are *isometric* but not necessarily based on icosahedral symmetry. Barley yellow dwarf and members of the luteovirus group may have octahedral particles.

Prunus necrotic ringspot, tobacco streak and other members of the isometric *labile* ringspot or ilar group of viruses (Fulton, 1968) contain RNA divided into pieces and encapsulated in different protein shells. Unlike the bromo- and cucumoviruses which have one size of capsid (isocapsidic viruses), the ilarviruses are heterocapsidic with two or three different-sized particles (Van Vloten-Doting *et al.*, 1977)—see 5.2.

4.2.3 Viruses with complex structure

This group includes cauliflower mosaic, geminate viruses, double-stranded RNA viruses (phytoreo- and fiji viruses), tomato spotted wilt, alfalfa mosaic, and rhabdoviruses. We shall now consider each briefly.

Cauliflower mosaic (CaMV). This is a double-stranded DNA virus with isometric particles 50 nm in diameter. Two coat proteins can be isolated, of MW 64 and 37×10^6, which together make up 84 % of the particle weight. The nucleic acid is thought to be protected by a double-layered capsid. The exterior capsid composed of the smaller protein may be in icosahedral configuration with $T = 7$, whereas the inner shell may be composed of the larger protein, and $T = 1$.

Viruses related to CaMV include dahlia mosaic, carnation etched ring and strawberry vein banding viruses.

Geminate viruses. Those readers who know their astronomy and/or astrology will guess that geminate viruses go about in pairs like the twins (gemini), Castor and Pollux. Viruses such as maize streak, beet curly top, bean golden mosaic, chloris striate and probably abutilon mosaic (Jeske and Werz, 1980) regularly show particles in pairs (Fig. 4.16). These are all single-stranded DNA viruses.

The detailed structure is known for chloris striate mosaic where the geminate particles are about 18×30 nm. Each member of the 'pair' is an incomplete icosahedron with $T = 1$ lattice, giving 22 capsomeres. The DNA in the pair is circular (Francki *et al.*, 1980) and it is suggested that 'paired' particles are required for infection, although both single and paired particles of beet curly top virus are infectious (Egbert *et al.*, 1976).

The DNA of bean golden mosaic virus is relatively small, with an MW of 7.5×10^5, and although the predominance (90%) of particles contain circular DNA, 10% contain linear molecules (Goodman *et al.*, 1980). These latter molecules may represent precursors of normal circular DNA that are packaged by mistake during virus assembly.

Double-stranded RNA plant viruses. These include rice dwarf and wound tumour viruses, members of the plant (phyto) *reovirus* group, and maize rough dwarf, oat sterile dwarf, pangola stunt, rice black streaked dwarf and sugarcane fiji disease viruses in the *fiji virus* group.

Plant reovirus resembles animal reoviruses in that the double-stranded RNA is divided into several pieces. The name derives from the fact that animal viruses in this group are associated with *r*espiratory and *e*nteric tissues but are *o*rphan in that they are not necessarily associated with a particular disease.

Rice dwarf and wound tumour viruses each appear to consist of 32 morphological proteins giving particles of 70 nm diameter. The RNA consists of 12 segments of different lengths. Both these viruses may have an outer membrane.

Fiji viruses have particles of 65–80 nm diameter, with a double protein shell and RNA in 10 segments. Both inner and outer protein shells are spiked (Fig. 4.16). The RNA of oat sterile dwarf has molecular weights of 2.76, 2.48, 2.35, 2.35, 2.08, 1.88, 1.18, 1.17, 1.16 and 1.00×10^6.

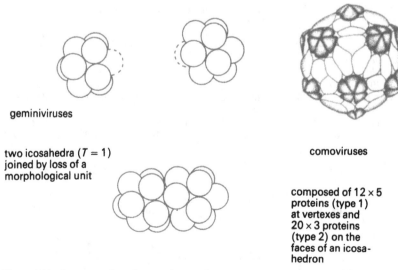

geminiviruses

two icosahedra ($T = 1$)
joined by loss of a
morphological unit

comoviruses

composed of 12 × 5
proteins (type 1)
at vertexes and
20 × 3 proteins
(type 2) on the
faces of an icosa-
hedron

Figure 4.16 Structure of geminate and comoviruses.

The outer coat of maize rough dwarf consists of 92 capsomeres plus 12 spikes (*A* spikes) at each icosahedron vertex. The inner layer possesses 12 *B* spikes coaxial with the *A* spikes. Oat sterile dwarf has a similar spiked construction.

Tomato spotted wilt (*TSW*). This virus is of interest since it has a membrane-bound isometric particle similar to influenza virus. Because this virus is unstable and there are many different isolates, the precise structure of the virus is not known. The virions contain 20% lipid, 7% carbohydrate and 5% RNA. The particles are 70–90 nm in diameter and the outer membrane consists of a nearly continuous layer of projections about 5 nm thick. Samples of TSW and influenza virus prepared for electron microscopy by similar procedures are almost identical in size, shape and internal appearance (Milne, 1967).

Alfalfa mosaic virus. Preparations of this virus yield particles of six different shapes and sizes. These are named bottom (*B*), middle (*M*) and top 'b' (*T*_b) components after position in the centrifuge tube on precipitation. The top component is further divided into *T*_a, *T*_o and *T*_z (the details of the components are given in Table 4.8).

The smaller components are spherical and may be icosahedral, and the larger elongated particles consist of two half icosahedra joined by a tubular net. Although the ends have five-fold axes, the tube has six-fold

Table 4.8 Details of the components in a preparation of alfalfa mosaic V.

Component	T_z	T_o	T_a	T_b	M	B
Particle weight ($\times 10^6$)	3.8	—	3.8	4.3	5.2	6.9
RNA type	$2 \times$ RNA-4	(*x*-RNA) (*z*-RNA)	$2 \times$ RNA-4	RNA-3	RNA-2	RNA-1
% weight	15.2–15.6		15.2–15.6	15.5	15.5	16.3
RNA MW	283×10^3	—	283×10^3	0.62×10^6	0.73	1.04×10^6
No. of nucleotides	883	—	883	1950	2250	3250
No. of protein subunits*	(60?)	(98)	132	150	186	240
Length of particles (nm)	—	—	30	35	43	56

* Nos. of subunits $= 60 + (n \times 18)$.

axes. The number of subunits in the capsids is 60 (for the two half icosahedra) + $18n$ where n is 10, 7, 5 or 4. The smallest particles may have smaller n-values (Heijtink et al., 1977).

The protein in all components is the same, consisting of 220 amino acids. The N-terminal amino acid is acetylated.

The RNA of the virus is divided into four species (RNA-1 to RNA-4). Intermediate in MW between RNA-3 and RNA-4 is x-RNA, and another smaller RNA (z-RNA) is also found and may be packaged in particles T_0 and T_3 and also in T_b and T_a.

Plant rhabdovirus group. This is a very interesting group of bacilliform or bullet-shaped viruses which are very similar to the animal viruses vesicular stomatitis (VSV) and rabies (RV). The animal viruses are called rhabdoviruses from the Greek rhabdos meaning rod. The plant viruses multiply in their leafhopper vectors, just as many of the animal viruses in this group have Diptera (flies) as alternative hosts. These are the most complex of plant viruses in chemistry and architecture; Table 4.9 lists a few plant rhabdoviruses and compares them with VSV and RV.

If we consider the example of sonchus yellow net (SYNV) in more detail, we find that chemically this, like other rhabdoviruses, consists of single-stranded RNA, MW 44×10^6, contained in a particle together with carbohydrate and probably some lipid. The proteins of the shell can be separated into glycoprotein (G), nucleic acid associated proteins (N) and

Table 4.9 Comparison of some features of plant and animal rhabdoviruses.

	PYDV	SYNV	AWSMV	VSV	RV
Shape	Bac	Bac	Bac & Bull.	Bull.	Bull.
Size—length	380	248	240–250*	175	180
			(200–215)		
			(405–411)		
width	75	94	75	68	75
Periodicity	5.0	41	—	4.6	4.5
Projections	√	√	√	√	√
Buoyant density	1.17	1.18	—	1.18	1.2
	810–950	1050	900	625	600
RNA MW	4.3×10^6	4.4×10^6	—	4.6×10^6	4.6×10^6
RNA %	—	<1	5	3	3

(PYDV = potato yellow dwarf; SYNV = sonchus yellow net; AWSMV = American wheat striate mosaic; VSV = vesicular stomatitis; RV = rabies.)
   * = range of particle sizes
   Bac = bacilliform
   Bull. = bullet shaped.

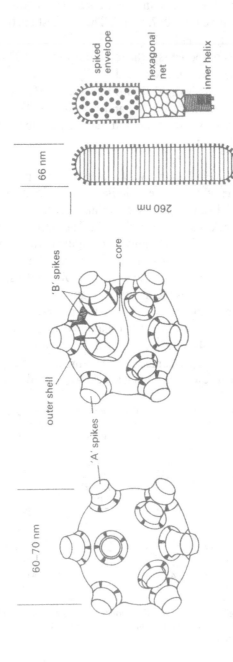

Figure 4.17   Structure of plant fiji, reo- and rhabdoviruses. Left, fiji disease virus—on right, part of outer shell removed to show core. Fiji and reoviruses have similar morphology (based on Hatta and Francki, 1977). Right, plant rhabdovirus, e.g. broccoli necrotic yellows virus—on right, layers removed to show internal structure.

Table 4.10 Comparison of molecular weights ($\times 10^3$) of proteins from plant and animal rhabdoviruses.

Protein	PYDV	SYNV	VSV	RV
L	?	?	190	?
G	78	77	69	78
N	56	64	50	58
M_1	33	45	29	35
M_2	22	39	—	22

L = protein associated with nucleocapsid
G = glycoprotein of projections
N = nucleoprotein
M_1 and M_2 = matrix proteins
PYDV = potato yellow dwarf
SYNV = sonchus yellow net
VSV = vesicular stomatitis
RV = rabies.

at least two membrane matrix proteins (M1 and M2). Small particles of other proteins are found consistently in purified preparations, and also an RNA-dependent RNA polymerase enzyme. The four major proteins (G, N, M1 and M2) have molecular weights of 77, 64, 45 and 39×10^3 respectively (Table 4.10).

Structurally these proteins make up a several-layered envelope. Electromicrographs show that the bullet-shaped particles are covered by projections. The envelopes of BNYV, SYNV and LNYV have been studied in particular detail and it appears that the envelope is composed of subunits arranged in a hexagonal lattice, and the projections are also arranged hexagonally (Fig. 4.16). The hexameric arrangement is skewed in relation to the longitudinal axis of the particle.

Internal to this envelope is a cylindrical structure which is striated, and can be seen in electron micrographs showing through the outer layer, giving the virus its characteristic appearance. This structure is most probably the nucleoprotein helix.

Bullet-shaped particles may be derived from bacilliform types either by breaking off of a terminal portion or as a result of incomplete assembly.

CHAPTER FIVE

PLANT VIRUS REPLICATION

5.1 Replication strategy

The processes involved in plant virus replication may include

(1) passage of virus through the cell wall;
(2) entry of virus or its nucleic acid into cells and then to replicative sites in cells;
(3) removal of protein from nucleic acid, this being termed 'uncoating'.

These early stages of infection are followed by

(4) translation of viral genome into replicase or a portion of that enzyme;
(5) replication of viral nucleic acid;
(6) replication of coat protein;
(7) assembly of new virus.

During, and resulting from, these events there arise symptoms of disease. Replication of plant viruses has been studied using whole plants or leaves, although many rapid advances have been made more recently using isolated protoplasts, i.e. plant cells with their walls gently removed by suitable treatment with cellulase and other enzymes. Some events thought to occur in plants have been postulated from events known to occur during bacteriophage and animal virus replication.

5.1.1 Early events

In many natural infections of plants, virus is placed directly into cells by vectors, so that early infection processes, (1) and (2) above, are completed by the transmission agent. In mechanical transmission, plant cell walls are broken and virus enters exposed protoplasts. Ectodesmata, (Fig. 5.1) cytoplasmic extensions through the outer wall of leaf epidermal cells, may be routes for virus uptake (Brants, 1964; Thomas and Fulton, 1968), but

94

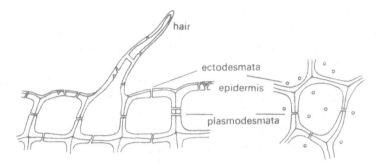

Figure 5.1 Ectodesmata—possible routes for virus entry.

have not been identified as paths of virus uptake in electron microscope studies (Merkens *et al.*, 1972). Furthermore, virus placed on undamaged leaves does not cause infection. On the other hand, virus can enter protoplasts without the need for them to be mechanically wounded (Takebe and Otsuki, 1969).

The entry of viruses into protoplasts may be by pinocytosis (Cocking, 1966; Cocking and Pojnar, 1969). More recently, Watts *et al.* (1980) have suggested that uptake by protoplasts comes about by a charge-dependent, temperature-independent process and that conventional pinocytosis is not involved. Some viruses, such as brome mosaic and pea enation mosaic, are positively charged and attach to protoplasts. Cowpea chlorotic mottle and TMV are negatively charged, and their entry into protoplasts is assisted by using positively charged molecules such as poly-L-ornithine.

At some stage in the infection process virus is *uncoated* by removal of protein from the nucleic acid. Evidence for uncoating has come from experiments showing that RNA derived from TNV gives local lesions on French beans quicker than whole virus particles. Also, UV light, which destroys nucleic acid, prevents infection if leaves are irradiated soon after inoculation. Leaves inoculated with TNV RNA quickly become resistant to UV irradiation, suggesting rapid production of intact virus, whereas leaves inoculated with whole virus are susceptible to UV treatment for a long period during which it is assumed virus is uncoated; only later is intact resistant virus formed.

Irradiation of PVX particles with UV light also lends support to the idea that early stages of infection involve uncoating. If PVX is partially inactivated by UV light and then exposed to daylight, there is some restoration of activity—photoreactivation takes place. But virus is not photoreactivable until 15–120 minutes after inoculation, and does not remain photoreactivable for more than three hours. If uncoated nucleic acid only is photoreactivable, then these results indicate that uncoating takes at least 15 minutes after inoculation and that new virus appears after three hours.

Bawden and Harrison (1955) and Niblett (1975) reported that, following mechanical inoculation, plant viruses became more sensitive to ribonuclease (RNase) and this may indicate the exposure of nucleic acid to enzyme attack by the uncoating process.

There is some evidence to suggest that uncoating occurs following attachment of virus to cell walls. Several workers have observed end-on attachment of rod-shaped viruses to cell walls following mechanical inoculation (Gerola et al., 1969; Merkens et al., 1972).

Gaard and DeZoeten (1979) observed this phenomenon in leaves infiltrated with virus. Using tobacco rattle virus, they showed not only that virus attaches end-on, but that there is gradual reduction in virus length; this was interpreted as uncoating. Kassanis and Kenton (1978) noted that TMV attaches and uncoats on leaf surfaces and in intercellular spaces. Cocking & Pojnar (1969) suggest that TMV particles become thinner as protein is digested from their surface, and that this happens to viruses within pinocytotic vesicles. Uncoating may, however, proceed in stages, since Shaw (1969) showed that the release of protein subunits from TMV is initially unaffected by low temperature and by cycloheximide, although later uncoating is influenced by these factors. Thus the early stages may be purely physical and later stages enzymic. In the isometric viruses turnip yellow mosaic (TYMV) and barley mosaic (BMV), most uncoating occurs in the first 10 minutes after inoculation. There may be a variety of mechanisms of uncoating, since TYMV results in empty protein shells, whereas BMV yields low molecular weight products (Kurtz-Fritsch and Hirth, 1972). Uncoating in these cases comes about by virus binding to cellular membranes. It is possible that the binding of virus, either to membranes or to cell walls, initiates the physical uncoating of virus which then triggers enzymic uncoating. Whatever the process, subtle changes in the coat protein may stop uncoating, as shown by Bancroft et al. (1971). Using cowpea chlorotic mottle virus, these workers found that replacement of arginine by cysteine in the coat protein resulted in loss of ability to uncoat and hence replicate. How this comes about is not understood.

5.1.2   Virus nucleic acid and protein synthesis

Having entered the cell and uncoated, the viral nucleic acid replicates, and proteins are produced, some enzymic, some viral and others of unknown

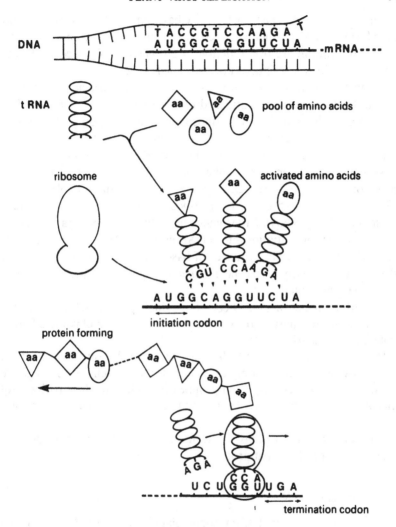

Figure 5.2 Schematic representation of protein synthesis.

function. Before considering these events, a brief resume of nucleic acid replication and the process of protein synthesis will be helpful. Fig. 5.2 shows diagrammatically the processes involved.

The base sequence of one nucleic acid can be used to make a complementary copy, or, in the case of DNA, may be used to make an RNA. During protein synthesis, the RNA complementary to the DNA strand contains the information for making protein. This RNA is called messenger or mRNA. The transfer of the base sequence of one nucleic acid in the formation of another is termed *transcription*. The enzymes involved in the transcription are *transcriptases* or *polymerases*. Virus-coded RNA-dependent RNA transcriptases are also called *replicases*.

The mRNA is *translated* into a sequence of amino acids to make a protein. Each amino acid is attached to a specific transfer or tRNA. The tRNA moves the amino acid to ribosomes, and aided by proteinaceous *initiation factors*, interpret the mRNA, so linking different amino acids together in a specific order depending on the type of protein being constructed. Amino acids are linked in a sequence determined by each triplet of tRNA bases recognizing the complementary triplet on the mRNA. Reading is carried out by the ribosomes which themselves contain ribosomal RNA (rRNA). Ribosomes bind and commence translation at specific translation initiation sites (TIS) on the mRNA. The amino acids are linked until a specific sequence of bases—the *stop codon*—is reached which, together with a specific *termination* factor, completes the construction of the protein.

A length of nucleic acid may code for a single protein, when it is called *monocistronic*, or it may be responsible for the formation of many different proteins (*polycistronic*).

All viruses synthesize an mRNA, and this is called positive strand RNA or +RNA; it may form complementary strands which will be negative RNA. Animal viruses have been grouped according to their pathway of mRNA synthesis (Baltimore, 1971), and this system may be used for plant viruses (see Fig. 4.1).

Formation of virus will involve transcription of viral mRNA and then translation to produce coat proteins.

5.1.3 *Replicative RNA*

When RNA is extracted from some plant cells infected with single-stranded virus, a double-stranded RNA fraction not present in healthy cells is found. This RNA consists of *replicative form* (RF) RNA which is unchanged by mild treatment with RNase, suggesting it is double-stranded. A second fraction, named *replicative intermediate* (RI) RNA, which can be modified by RNase treatment, is also present. RI consists of double-stranded RNA with single-stranded 'tails'. The precise structure of RI is still controversial, but it may consist of a full-length parental negative RNA strand with 6–8 daughter positive strands at stages in the process of synthesis (Fig. 5.3).

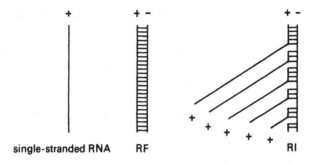

single-stranded RNA   RF   RI

Figure 5.3   The structure of replicative form (RF) and replicative intermediate (RI) forms of RNA.

Evidence for the production of RF and RI forms of RNA has come from the study of a number of viruses, including TMV, tobacco ringspot, southern bean mosaic, plant tombus viruses and TNV. RF of TMV is not infectious, but becomes so when the strands are separated (Jackson *et al.*, 1971); multipartite viruses such as cowpea mosaic and alfalfa mosaic produce multiple forms of RF. There is considerable evidence that RF and RI are intermediates in the replicative cycle of positive-strand RNA plant virus (Siegl and Hariharasubramanian, 1974).

The replication of plant viruses will now be considered, for convenience under two headings, replication of RNA viruses, and replication of DNA viruses.

5.2 Replication of RNA viruses

RNA viruses may contain (1) positive strand RNA, (2) negative strand RNA, or (3) double-stranded RNA. Let us look at the replication of viruses in each of the RNA groups listed above.

5.2.1 *Positive strand RNA viruses*

This is by far the largest group of plant viruses. In these viruses, the viral RNA acts as a template for synthesis of negative strands, with the formation of RI and RF forms of RNA. The negative strand gives rise to new daughter positive RNA. The synthesis of negative strand RNA must involve the activity of RNA polymerase which may arise either (*a*) by activation or modification of host polymerase or (*b*) by formation of viral polymerase. In the latter case, viral positive RNA must be translated to give protein with polymerase activity; the viral+RNA acts therefore as mRNA.

Virus RNA is also 'genome' RNA since it contains genetic information for virus protein replication, for the enzymes necessary for replication and also for properties such as symptom production and vector specificity.

Translation of viral RNA. Studies of the translation of plant virus mRNA show three methods of genome expression.

1. The RNA may be polycistronic, and contain a number of translation initiation sites (TIS) each corresponding to a gene. Each TIS is recognized by ribosomes and each gives rise to a protein. Thus TNV, to give only one example, can be translated into three types of polypeptide from three independent translation initiation sites.

2. Genome expression may be pseudomonocistronic, the genes being read by ribosomes proceeding from 5' to 3' terminus of the mRNA to produce a single large protein, which is subsequently cleaved into smaller virus-specific proteins. This continuous translation is best understood for animal picorna viruses, but como- and nepo- plant viruses are also translated in this way (see multipartite viruses).

3. Virion RNA may contain genes that are not immediately translatable, and such 'closed' genes are not accessible to translation by ribosomes to make proteins. Closed genes can be activated by 'autonomization' (Atabekov and Morozov, 1979) into a separate mRNA.

With the common strain of TMV, for example, Hunter *et al.* (1976) showed that ribosomes can translate cistrons for 140 000 and 160 000 molecular weight proteins. The sum of these molecular weights is too large to be coded separately and their genes must overlap. This was substantiated when removal of small portions of the 5' end of TMV RNA resulted in failure to produce either protein. Surprisingly, no coat protein could be detected when TMV RNA was translated in *in vitro* cell-free protein synthesizing systems from *E. coli*, reticulocytes or wheat germ.

Samples of TMV RNA could produce coat protein *in vivo*, so it could be concluded that the genomic TMV RNA is not an efficient template for coat protein translation in *in vitro* experiments. In later experiments it was shown that TMV induced smaller-than-genomic RNA in tobacco tissue and in protoplasts. This low molecular weight RNA (LMC RNA) has an MW of about 280 000 and is derived by transcription from the 3' end of the TMV RNA. The LMC RNA acts as an efficient mRNA for coat protein production (Fig. 5.4). In the cowpea strain of TMV, but not the common strains, LMC RNA is separately encapsidated to produce short rods about

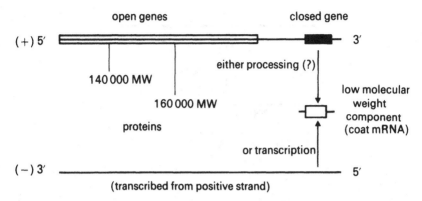

Figure 5.4 Map of TMV RNA, showing possible course of production of coat mRNA (after Hunter *et al.*, 1976).

40 nm long. In addition, the cowpea strain, and some other strains, produce intermediate rods shorter than 300 nm, which contain mRNA (I_2 RNA) for the coat protein and for a protein of MW 30 000 with unknown functions.

Multipartite viruses. As mentioned previously, some viruses require more than one type of particle to be present in their host before replication can take place. This is because their genome is divided between different pieces of nucleic acid, each of which is encapsidated separately.

Positive strand plant viruses may be divided into two- and three-component viruses.

Two-component systems. These are found in comoviruses, nepoviruses and tobraviruses. Cowpea mosaic virus is the type member of the *comovirus group*. It has a single stranded RNA genome divided within two spherical particles 28 nm in diameter. Both types of particle or their nucleic acid are required for replication. The two RNAs have molecular weights of 2.02×10^6 for the bottom (*B*) component, and 1.37×10^6 for the middle (*M*) component. Top components contain no RNA. Each has 3' polyA ends (El Manna and Bruening, 1973) and a covalently-bound protein of about 5000 MW attached at their 5' end (Daubert *et al.*, 1978). The *M* component RNA is thought to code for coat protein, and the *B* component RNA induces production of proteins, some with replicase activity. Recent work (Razelman *et al.*, 1980) has suggested that the *B*-component RNA is translated into a single 200 000 MW protein which is cleaved to produce smaller proteins (Fig. 5.5).

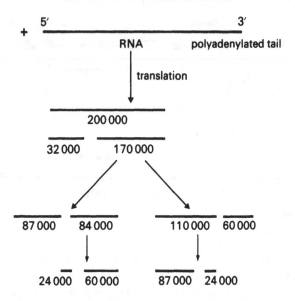

Figure 5.5   Model for the expression of cowpea mosaic virus RNA (bottom component), and the subsequent cleavage of proteins.

The *nepoviruses* separate on centrifugation into top component devoid of RNA, together with middle (*M*) and bottom (*B*) components that contain RNA of different molecular weights. Two RNAs are found, RNA-1 with molecular weight about 2.4×10^6, and RNA-2 with molecular weight in the range $1.4 - 2.2 \times 10^6$. Bottom components contain either one molecule of RNA-1 or two molecules of RNA-2, and middle components contain one molecule of RNA-2 depending on type of virus. Both RNA-1 and RNA-2 are required to produce infection. In the type member, tobacco ringspot virus, RNA-1 carries determinants for host range and seed transmissibility, whereas RNA-2 determines serological specificity and nematode transmissibility. With raspberry ringspot and tomato black ring viruses in *Chenopodium quinoa*, RNA-2 controls lesion production, whereas RNA-1 and 2 determine speed of spread of disease through these plants.

The type member of the *tobravirus* group is tobacco rattle virus (TRV). Tobraviruses are nematode-transmitted, like the nepoviruses, but are rod-shaped. Examination of TRV samples by electron microscopy shows mixtures of long particles (180–210 nm) and short particles (45–115 nm). Long particles can infect intact plants and protoplasts, but virus particles cannot be detected in the sap. Long particles apparently have the RNA replicase genome but lack the coat protein gene. Short particles contain the coat protein determinants and are necessary for formation of complete virions. This situation is similar to the requirement by cowpea strain of TMV for the LMC RNA which is encapsidated into shorter than normal rods (see p. 100).

In more recent studies of TRV (Pelham, 1979; Bisaro and Siegel, 1980) a third RNA of MW 550 000 has been demonstrated. This RNA-3 may be packaged separately in rods about 43 nm long. The production of this additional RNA probably comes about by transcription from the 3' end of RNA-1, and is comparable but not identical to the replication strategies of TMV (see Fig. 5.4). TRV may therefore be a three-component virus like those described below.

Three-component virus systems. Viruses with three RNA species in separate capsids, each required for infection, have been grouped into two major sections, *isocapsidic* and *heterocapsidic* types, by Van Vloten-Doting *et al.* (1977).

Isocapsidic viruses have separate genomic RNAs, encapsidated into identical capsids. In the second group, the heterocapsidic viruses, separate RNA species are encapsidated into particles of different dimensions dependent on RNA size. In the case of alfalfa mosaic, particles vary from small spheres to larger bacilliform shapes.

Viruses with tripartite genomes often contain more than three RNA species. One in particular, a small RNA of 3×10^5 MW (RNA-4), occurs in appreciable amounts and can be encapsidated alone or together with the smallest genome RNA. In the case of the isocapsidic bromo- and cucumoviruses, RNA 1, 2 and 3 only are required for infection. With ilarviruses and alfalfa MV, infectivity is lost without RNA-4. Infectivity is restored not only by RNA-4 however, but also by addition of a small amount of its translation product, the coat protein. Ilarviruses and alfalfa MV are therefore 'protein dependent' and brome- and cucumoviruses 'protein-independent' tripartite genome viruses.

Brome mosaic virus (BMV), an isocapsidic tripartite genome virus, contains four species of RNA designated 1, 2, 3 and 4, of molecular weight 1.1, 1.0, 0.75 and 0.3×10^6 respectively. RNA-1 and 2 are separately encapsidated, whilst RNA-3 and 4 occur together. Replicative intermediates of RNA-1, 2 and 3 are formed but there is disagreement concerning the existence of RI corresponding to RNA-4. Each RNA species is mono-cistronic in cell-free protein synthesis systems and is translated into single polypeptides of molecular weight 120, 110, 35 and 20×10^3 respectively, the latter being the coat protein. Such polypeptides are formed in BMV-treated protoplasts from systemic, local lesion and non-host plants (Okuno and Furusawa, 1979).

As in bromoviruses, four major RNA species are to be found in the cucumoviruses. RNA-1 (1.3×10^6 MW), and RNA-2 (1.0×10^6 MW) occur in separate particles whilst RNA-3 (0.80×10^6 MW) and RNA-4 (0.33×10^6 MW) are encapsidated together. The coat protein genes are located with RNA-3, as well as genes determining transmissibility by aphids. RNA-4 is monocistronic, containing the coat protein gene, and is probably derived from RNA-3. All four RNA types can be charged at their 3′ end with tyrosine, using amino-acyl tRNA synthetase, and their 5′ ends are capped with 7-methyl guanosine. Replicative form RNA has been found for each genomic RNA in studies using protoplasts.

The most intensively studied of the viruses in the heterocapsidic tripartite genome group is alfalfa mosaic (AMV). The four genomic RNAs are packaged separately. The largest RNAs(1–3) of molecular weights 1.04, 0.73, 0.62×10^6, comprise the viral genome. RNA-4 (MW 0.28×10^6) is a subgenomic messenger for the coat protein. RNA-1 and 2 are mono-cistronic but RNA-3 is bicistronic, coding for the coat protein (24 280 MW) and also for a larger protein of 35 000 MW. Genes for symptom production are carried on RNA-1 and 2.

Double-stranded RNA corresponding to each of the genome RNA species have been isolated from infected plants. Replicative forms of RNA corresponding to RNA-4 have not been detected. The 5′ ends of all RNAs have a 5′ blocked methylated cap, no polyA tails and cannot be charged with amino acids at their 3′ ends.

Ilarviruses, such as tobacco streak, citrus leaf rugose and citrus variegation viruses, resemble AMV in that they are heterocapsidic with tripartite genomes and also protein-dependent. No serological relationship, however, has been found between AMV and other tripartite viruses. It seems remarkable then that proteins from AMV and from ilarviruses can reciprocally activate each other's genome to produce infectivity.

It can be seen from these accounts of positive-strand RNA viruses that replication varies widely in complexity. Before leaving this group of viruses, a brief consideration of satellite viruses seems appropriate.

Satellite viruses and satellitism. Tobacco necrosis virus preparations contain particles of 17 nm diameter as well as the normal 26 nm diameter particles. These smaller particles can be isolated from TNV by extracting leaves at pH 4.5, under which conditions the TNV is less soluble than the

Table 5.1 Comparison of defective interfering particles (DI) and TNV satellite virus (STNV).

	DI	STNV
Replication	Requires normal virus	Requires TNV
Interference	Yes	Yes
Capsid	Normal virus	Different sizes and serologically unrelated
Genome	Part of normal virus	Less than 30 nucleotide homologies

smaller viruses. These small viruses are called *satellite viruses* (STNV) and have been found to replicate only in the presence of TNV. Some 20 % of the STNV particle weight is made up of single-stranded RNA of MW 0.28×10^6. This RNA is sufficient to code for coat protein but evidently not sufficient to bring about complete replication. Other small isometric viruses cannot substitute for TNV.

Similar satellite systems have been described for tobacco ringspot, raspberry ringspot and myrobalan latent ringspot viruses in the nepovirus group.

The satellite associated with TNV has similarities to the *defective interfering particles* (DI) found associated with many groups of rapidly replicating animal viruses (Table 5.1), such as poliovirus. DI particles contain only part of the virus genome (subgenomic), and they hinder the production of normal viruses by competing for polymerases. In addition, DI viruses only replicate in the presence of the normal virus and they have the same coat protein and part of their genome in common with that of their normal 'helper' virus. Satellite viruses, however, are serologically unrelated to TNV and the homology between their RNAs is less than 30 nucleotides.

Tomato black ring (TBRV) and tobacco ringspot virus (TRSV)—both nepoviruses—behave much more like the DI system of animals, since their satellite particles require normal particles for replication and have coat proteins identical to the normal TBRV and TRSV particles.

Cucumber mosaic virus, as we have seen, has a multipartite genome with RNA divided into three genomic and one sub-genomic portion. An RNA, called CMV-associated RNA-5 (CARNA-5), is also to be found evenly distributed amongst CMV particles. CARNA-5 replication depends on CMV replication and is therefore considered as *satellite RNA*. No large nucleotide sequence homologies exist between CARNA-5 and genomic

RNAs. CARNA-5 has mRNA properties and directs the formation of two polypeptides of unknown function.

5.2.2 Negative strand RNA viruses

Negative strand RNA cannot act as a template for protein synthesis. Viruses with this type of nucleic acid (such as plant rhabdoviruses) need to replicate positive strand RNA which can act as an mRNA. Formation of positive strands from negative requires a virus-coded enzyme—an RNA transcriptase (polymerase). How then can these viruses utilize an enzyme that they cannot make? The solution to this problem is relatively simple. The virus-specific transcriptase formed in cells during infection is incorporated into the virus as part of its structure. On infection the virus introduces into its host a negative-strand RNA genome, together with the replicase enzyme required to produce positive-strand mRNA. Evidence that these events occur in plant viruses comes from studies of sonchus yellow net virus, where a positive strand RNA complementary to the virus RNA can be detected, and is associated with polyribosomes. The RNA-directed RNA polymerase which transcribes the single-stranded negative RNA genome of lettuce necrotic yellows virus has been detected in the virions, although efforts to detect similar enzymes in the virions of other plant viruses and in rabies virus itself have proved unsuccessful. The positive RNA transcribed contains a polyA sequence at its 3′ end, suggesting that it is an mRNA.

5.2.3 Double-stranded RNA viruses

Viruses in the plant reo- and fiji groups resemble animal reoviruses and contain double-stranded RNA in 10–12 segments. Replication mechanisms of plant and animal reoviruses are probably similar. Wound tumour virus (WTV), like animal reoviruses, contains virus-associated RNA polymerase and methylase enzyme activity (Rhodes et al., 1977). The latter enzyme occurs in the 5-capped terminal structure of the WTV mRNA. This cap resembles that found in the insect reovirus, cytoplasmic polyhedrosis virus, and in reovirus itself.

mRNA is formed by asymmetric transcription, that is, only one of the RNA strands in each of the double-stranded RNAs is copied. The transcriptase (polymerase) involved is found in the core of the virus. In animal reoviruses this transcriptase is activated by removal of part of the outer capsid. WTV transcriptase does not, however, need to be activated in this way (Black and Knight, 1970). Transcriptase remains attached within the virus core, and transcription occurs from the 10–12 genome segment by a conservative mechanism yielding mRNAs equivalent in size to the genome segments. Smaller RNAs are transcribed more often than larger-sized RNAs. Each mRNA is translated directly into protein.

The formation of progeny double-stranded RNA is quite different from the replication of double-stranded DNA. This is clear from the observations that parental reovirus RNA is not uncoated, and that genetic information is passed to progeny by positive strand mRNA. Such positive RNA has the dual role of (a) being translated into protein and (b) of acting as the template for negative-strand RNA production. Association of the negative strand with the positive strand results in double-stranded progeny RNA. Free negative-strand RNA is not found associated with reovirus replication, and negative strands are constructed in proteinaceous core structures similar to the core portion of mature virions. To form the pre-core, 10–12 single positive-stranded RNAs come together with the appropriate protein. How each pre-core comes to receive only one set of 10–12 positive RNAs is not understood.

5.3 Replication of DNA viruses

These are divided into *caulimoviruses*, such as cauliflower mosaic (CaMV) and dahlia mosaic with double-stranded DNA, and *geminiviruses*, such as bean golden virus, which contain single-stranded DNA.

5.3.1 *Caulimoviruses*

Most work on caulimovirus replications has been done on CaMV. The DNA is circular, but up to 10% may be linear—this may be a breakage product of the circular molecule (Volovitch *et al.*, 1978). An interesting feature of the DNA is the presence of a small (less than 1%) covalently-linked section of RNA (Hull and Shepherd, 1977). RNA sequences are present in the DNA of some animal viruses and possibly in eukaryotic nuclear DNA, but the function of the RNA segment is not known.

Only one strand of the viral genomic DNA is transcribed into RNA. Since polymerase activity cannot be demonstrated in CaMV particles, it is assumed that host cells provide this enzyme. This is not unusual, since small DNA bacteriophage and DNA animal viruses are dependent on their host polymerase (Shepherd, 1976). In recent studies, Covey and Hull (1981) have detected several RNAs transcribed from CaMV DNA in turnip leaves. The spectrum of RNAs varied with time, but the appearance time sequence has not been determined. An RNA compound of 2.3×10^3 nucleotides acts as an mRNA producing a 62 000 MW polypeptide. This mRNA has polyadenylated 3' which is probably added after transcription from the DNA.

5.3.2. Geminiviruses

These viruses have single-stranded DNA which may be linear but is more usually circular in form. Common nucleotide sequences between the two topological forms of DNA suggest that linear molecules are derived from circular molecules. The low molecular weight $(7-8 \times 10^5)$ of the DNA suggests that the virus may be multipartite, the genome being divided between different particles. Haber *et al.* (1981), using restriction endonuclease enzymes to determine nucleotide sequences, showed that in bean golden mosaic virus (BGMV) the nucleic acid composed of 2510 nucleotides behaved as if it were composed of 5000 nucleotides. These results support the idea that the genome is divided between two circular DNA molecules differing only in nucleotide sequence and hence genetic constitution.

Single-stranded DNA viruses from plants probably replicate in a fashion similar to that of mammalian parvoviruses, although the latter have single-stranded linear rather than ring form DNA. Plants infected with BGMV contain double-stranded DNA with one strand complementary to viral DNA. Geminivirus DNA synthesis probably depends on enzymes provided by the host since the information on the viral genome is very limited (Goodman, 1981). Transcription of the double-stranded DNA would presumably produce mRNA from which coat protein would be translated.

5.4 Assembly of plant viruses

Following the formation of viral nucleic acid and protein subunits, assembly into complete virions takes place. Helical viruses will be considered first and then isometric viruses.

5.4.1 Helical (elongated) viruses

Initially the RNA of elongated viruses may show some degree of secondary structure with 30–70 % base pairing, depending on the ionic strength of the

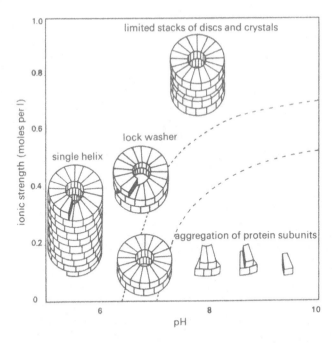

Figure 5.6   Effect of pH and ionic strength on the assembly of TMV protein subunits (based on Butler and Klug, 1978).

medium. Encapsidation involves the splitting or 'melting' of this RNA and coating with protein. In the case of papaya mosaic virus, the coat protein induces melting, although this is not the case with TMV and its coat protein (Erikson and Bancroft, 1981). Such melting of RNA seems to be a feature of helical virus assembly, since the RNAs of isometric viruses appear to undergo little change in secondary structure during maturation.

The assembly of TMV comes about by the aggregation of protein subunits (monomers) into trimers and pentamers (Fig. 5.6). Double discs are then formed which in turn form rods at near-neutral pH. Helical rods are formed at a pH below 6 in the presence of TMV RNA. The nucleic acids are threaded into discs as they pile up. It is claimed, however, that TMV assembly commences, not at the end of the RNA molecule, but at an internal region some 800–1000 nucleotides from the 3′ terminus, and proceeds in two directions. Viral rod elongation is rapid, reaching the 5′ end within 5–7 minutes, whilst elongation to the 3′ end is much slower (Fukada *et al.*, 1978). The internal initiation site for coat assembly is only about 300 nucleotides from the 3′ end in the cowpea strain of TMV. This is

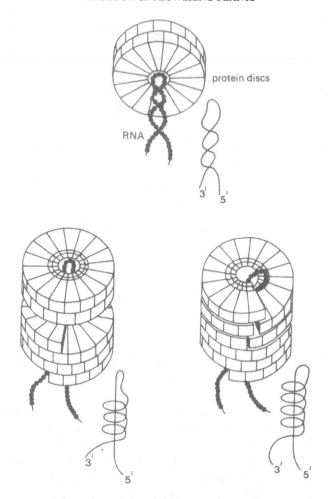

Figure 5.7   Nucleation and elongation of TMV (based on Butler and Klug, 1978).

within the coat protein cistron, and this explains why the coat protein mRNA of this strain is coated with protein to make short particles, whereas in the other strains it is not encapsidated.

Recently Bisaro and Siegel (1980) have suggested that a 30 000 MW polypeptide translated from RNA-3 of tobacco rattle virus may function in some unknown way in the assembly of this rod-shaped virion.

The length of rod-shaped viruses depends on the nucleic acid length. The helical arrangement of protein subunits follows the helical structure of nucleic acid. In rigid rods such as TMV, the protein subunits fit together in equivalent environments, but in flexible rods such as potato virus X and henbane mosaic virus, bonding between subunits is less well organized and involves some quasi-equivalent bonding (Martin, 1978). Assembly of these flexible virions involves assemblage of single-layer discs rather than double discs (Butler and Durham, 1977).

5.4.2 Isometric viruses

The assembly of isometric viruses involves the formation of aggregates of protein subunits into outer shells (capsids) enclosing nucleic acid. Many small isometric plant viruses produce empty shells during replication, indicating that, like the protein of TMV, isometric virus production involves self-assembly and strong protein interactions. Whether shells are made and then filled with nucleic acid is not known. However, whole virus particles of turnip yellow mosaic treated with 1 M KCl at pH 11.6 lose their RNA and leave empty shells, and five of the 180 protein subunits are lost with the RNA. This evidence suggests that assembly may be more complex than simply filling empty shells with RNA. Nucleic acid may complex with a few protein subunits at specific points, and this may lead to capsid formation by aggregation of additional proteins. There is little evidence to show that specificity exists between nucleic acid and protein in virion formation. Cowpea chlorotic mottle virus (CCMV), for example, can encapsulate the RNA from TMV, producing infectious particles. Furthermore, RNA of CCMV can be hybridized to protein of brome grass mosaic (BGMV). CCMV RNA can also be encapsidated by a mixture of CCMV and broad bean mosaic virus protein. It must be pointed out, however, that in general viral proteins do not combine with host nucleic acid, indicating some degree of specificity.

Because plant virus genomes contain more information than is required for specification of coat protein and a RNA replicase, it seems possible that additional proteins or polypeptides necessary for virus assembly may be produced. Such maturation proteins are known for bacteriophages (Steitz, 1968). Certainly evidence from studies of the multicomponent bromo-viruses indicates that products from RNA-1 or RNA-2 are required for the correct assembly of capsomeres formed by RNA-3, which contains the coat protein genome.

5.4.3. Membrane-bound viruses

The assembly of plant membrane-bound viruses such as reo- and rhabdo-viruses may occur, as with animal viruses, in three main steps: (1) the assembly of nucleocapsids; (2) the insertion of structural proteins into the host cell membranes; and (3) a budding process in which the nucleocapsids are enclosed by the modified membranes and released from organelles or cells. With rhabdovirus, for example, the outer membrane may be formed from the inner lamella of the nuclear envelope.

Sites of virus replication. Using a variety of techniques, including nucleic acid staining, phase contrast microscopy, UV light absorption and examination of radioactive nucleotides, it is possible to locate centres of virus replication. Evidence from these methods, together with observation by electron microscopy of the accumulation of virus particles and inclusions, suggests that nuclei, nucleoli and the cytoplasm are the main sites of virus replication, with some replication in chloroplasts and other organelles.

The nucleus is the site for replication of potexviruses, caulimoviruses, geminiviruses and tobamoviruses. An example from the latter group, TMV, is reported to accumulate in the nucleolus, move through the nucleus and into the cytoplasm. TMV replication, however, is inhibited by cycloheximide which stops cytoplasmic RNA translation. Chloramphenicol which prevents chloroplast RNA translation, has little effect on TMV replication. Rhabdoviruses, such as sowthistle yellow vein virus (SYVV) and maize mosaic (MMV), are usually located in the perinuclear

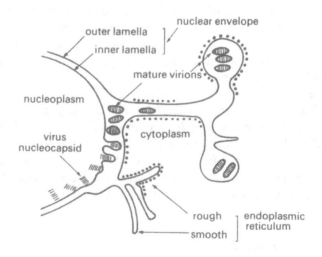

Figure 5.8 Diagram to show possible sequence of events and sites of assembly of plant rhabdoviruses (based on Francki, 1973).

space of host cells (Fig. 5.8). The probable site of nucleocapsid formation is within the nucleus; particles then acquire their outer membrane by budding from the inner nuclear membrane. Other rhabdoviruses, e.g. lettuce necrotic yellows, accumulate in cytoplasmic vesicles, particles having budded from the endoplasmic reticulum (Francki, 1973). Studies of sonchus yellow net virus show that mRNAs concerned with virus replication are partitioned in the cytoplasm, and possibly transported to the nucleus, where they function in viral morphogenesis.

Bromoviruses, and members of the poty-, nepo- and comovirus groups all accumulate in the cytoplasm of their host cells and probably replicate there rather than in the nucleus. Chloroplasts seem to provide the replication sites for tymoviruses, since peripheral vesicles formed by the invagination of the chloroplast membrane accumulate virus-related proteins and replicative forms of RNA. On the other hand, although tobraviruses are frequently found associated with mitochondria, it is not certain that these are sites for their replication (Harrison and Robinson, 1978).

Virus strains. As we have seen, viruses, like other living organisms, replicate by making copies of their nucleic acid. The nucleic acid in turn dictates the structure of their proteins and other structural components. Any alteration in the initial copying of the genome will lead to progeny complete but different from the parents. Gibbs and Harrison (1976) use the term *variant* to describe any novel virus isolate, and the term *strain* to describe naturally-occurring variants. Variants and strains are recognized by differences from the original isolate in terms of symptom expression, host range and means of transmission, as well as by biochemical features. Variants arising by mutations may be induced *experimentally* by treatment of virus with chemicals such as nitrous acid, the nucleic acid base analogue 5-fluorouracil and 8-azaguanine, and also by X-rays, UV light and elevated temperatures. Natural mutations producing new strains occurs at rates estimated at 1 in 1000 to 2 in 100. These figures record only mutations sufficient to produce recognizable symptom changes and many under-estimate actual mutations.

These authors place variants of plant viruses into one of three categories.

(a) Variants arising from alteration (by physical or chemical events) of a single parental genome. Alteration of nucleotide bases gives rise to *mutants* and loss of bases are *deletion mutants*. Nucleotide base changes result, amongst other things, in changes in coat protein amino acid composition.

(b) Where the genome is derived from more than one parental nucleic acid, the breakage and joining of new covalent linkages in the nucleic acid gives rise to variants by recombination, the progeny being *recombinants*. This type of variation arises particularly where the genome is divided between different pieces of nucleic acid.

In mixed infections of *two different viruses*, nucleic acids from each may reassort and segregate to give rise to new types of virus, these being called *pseudo-recombinants*.

(c) The third type of variant is where one virus is assisted in the production of its coat protein, for example by another variant of itself or another virus in a mixed infection. The new virus is said to have acquired its unique properties by *complementation*.

CHAPTER SIX

PLANT VIRUS DISEASE CONTROL

Many plant pathogens, particularly fungi, can be controlled by the application of chemicals which interfere in some way with the metabolism of the invading pathogen, and so prevent or ameliorate disease. Unfortunately, these methods cannot be used so extensively to control plant viruses. Having few, if any, enzymes of their own, viruses depend either on enzymes already in host cells or on those that are induced as a result of infection. These enzymes are responsible for nucleic acid and protein synthesis, and chemical interruption of their activity disrupts similar enzymes essential for the normal functioning of cells. Chemical attack on viruses often results in death of cells and tissues and possibly of whole plants. Control measures other than direct chemical attack on the viral pathogen must be attempted. A knowledge of the identity of the invading virus or viruses, the source of infection and the means of viral transmission, allows control measures to be formulated. Prevention, or at least alleviation, of the effects of viruses, involves:

(1) Elimination of sources of virus.
(2) Elimination of the virus from infected plants.
(3) Control of vectors.
(4) Breeding for resistance and the use of cross-protection methods.

Each of these approaches to control will be considered.

6.1 Sources of virus

Chapter 7 gives some information regarding the methods available for examining plant material in order to establish that observable symptoms result from virus infection. If possible, the identity of the virus should be

115

established; this may help in determining the source of infection. Possible sources include (a) weeds, (b) other crops, (c) debris and crop residues, (d) tools and personnel, (e) seeds and (f) diseased propagation stock plants. Elimination of these sources of virus will effect some control of disease. Let us look at each in turn.

6.1.1   Weeds

Weeds may act as reservoirs of virus or of vectors, within the crop or in areas adjacent to crop plants.

Virus-infected weeds within a crop are a serious hazard because of the close proximity of the contaminated plants to the healthy ones. Control can be achieved by elimination of the weeds by 'roguing out' or by using herbicides. Control by such measures may not be easy since virus may be present in the seeds of weeds. For example, cucumber mosaic virus and some nematode-transmitted viruses may be spread through the seeds of the annual *Stellaria media* (chickweed). Perennial weeds may act as sources of virus in diseases of legumes and cucurbits (Thresh, 1978). Nearby wild or garden sources of infection are important in the epidemiology of virus diseases of celery, pepper, potato and lettuce (Duffus, 1971; Thresh, 1976).

6.1.2. Other crops

Disease may enter plants from similar crops in adjacent fields, or from distant ones if mobile vectors are involved. In other cases, virus from one type of crop may spread to another as, for example, the spread of barley yellow dwarf virus between red clover and field beans. The production of field crops in overlapping sequences, as in the case of lettuce, brassicas and cereals, may result in the spread of virus from old to new plantings.

Ornamentals in glasshouses are offered some protection from outside invasion by vectors. Once virus-carrying vectors have entered glasshouse crops, cultural conditions are such that disease is intensified and spreads rapidly. Ornamentals may become infected with viruses from agricultural crops, but in general the reverse is not true, and ornamentals offer very little danger as virus sources for agricultural crops (Hollings and Stone, 1980).

6.1.3   Debris and crop residues

Crop plants are rarely entirely harvested, and there remains behind some remnant of the plants which may contain virus. In most cases the nonuseful portion of the crop is removed later or burnt *in situ*. In addition, some useful parts may fail to be harvested entirely, for example 'volunteer' potatoes may be left buried in the soil, and these may act as sources of virus for subsequent plantings of the same crop. Rotation of crops helps to prevent this. Debris which does remain should, if possible, be removed and burnt. The removal of debris and careful crop hygiene are particularly important with glasshouse crops—pots, tools, and benches should, as far as is practicable, be sterilized and kept clean and in good order.

6.1.4 *Tools and personnel*

Virus may be spread by direct contact with diseased plants or with contaminated tools, machinery and clothes. Control in these cases involves crop hygiene.

Minimal contact should be made by workers, their clothes and tools with plants, and particularly with infected ones. Hands and tools can be decontaminated using trisodium orthophosphate (3–5%). Smoking should be avoided in nurseries, particularly when handling tomato plants, since TMV is present even in cured tobacco and is passed from the hands of workers to plants. Contact spread is highest in diseases of glasshouse crops such as tomato, cucumber, chrysanthemums and carnations where plants are frequently handled routinely. The use of pelleted seeds helps in this respect, since seeds can be spaced out and the need for 'pricking out' minimized.

Pruning of plants should be done carefully and the pruning knife should initially be heat sterilized and subsequently decontaminated by dipping into trisodium orthophosphate (Broadbent, 1964).

6.1.5 *Seeds*

Viruses transmitted by nematodes are also frequently transmitted by seeds; other seed-transmitted viruses are listed in Table 3.1. Although virus is contained within the seed, the pathogen can be eliminated by treatment of the dormant seed by heat or by irradiation with UV light. These procedures do not always inactivate virus without damage to the embryo (Slack and Shepherd, 1975). Another approach is to treat germinating rather than dormant seeds. In experiments described by Cooper and Walkey (1978), seeds and embryos of *Nicotiana rustica* infected with cherry leaf roll virus remained infected after culture at 22°C, but no infectivity could be detected after five days at 32°C. Infectivity could be found when such cultures were returned to 25°C for eight days. Permanent eradication of virus required seven days' incubation at 40°C.

6.1.6   Propagation stock plants

Where propagation of plants is by means of vegetative parts, care must be taken to ensure that virus-free material is used. It is important, therefore, that rootstocks, scions, budwood, runners, rhizomes, bulbs, corms, tubers and other vegetative parts used for propagation are *indexed* to show that they contain no viruses, or at least none of economic importance.

Virus indexing—visual inspection of plants for virus symptoms—may not always be possible or reliable. Seeds and deciduous woody plants may not show external symptoms and this makes it difficult to select virus-free material for propagation. A number of tests are carried out, either singly or in combination, to index plants. Ideal tests are reliable, rapid, cheap, specific, and if possible simple to carry out. Such tests should also be suitable for routine testing of large numbers of samples.

The most commonly used tests include:

1. *Infectivity testing*—this involves either (*a*) inoculation of crude extracts to a host or hosts that give distinctive symptoms, or (*b*) grafting suspected material on to sensitive hosts.
   These techniques are slow and it is often difficult to find reliable means of transmission or suitable hosts.
2. *Serology*—using antisera produced against known viruses it is possible to test sap or crude extracts cheaply and rapidly with high specificity. Various tests are used based on precipitin reaction (see section 7.4). Recently the ELISA technique (section 7.4.2) has been developed and used extensively to test for a wide variety of plant viruses.
3. *Electron microscopy*—this can be used to determine if virus particles are present, and their shape. It is not always possible, however, to obtain reliable results, particularly if virus particles are present in low numbers. Improvements to the reliability and sensitivity of the method have come about recently by combining electron microscopy with serology, where observation is made of the specific binding of virus to antigen-coated electron microscope grids. There are several applications of the technique (see section 7.4.3).

Using these indexing techniques it is possible to test propagative tissues and show whether they contain virus. If disease-free plants are obtainable they can be used as stock plants from which more virus-free plants can be propagated (see section 6.2.4).

6.2 Elimination of viruses from plants

The second major approach to virus disease control can be thought of as methods of eliminating virus from infected plants. Three principal techniques are involved: chemotherapy; thermotherapy, and meristem culture. For the most part these methods are designed to produce virus-free stock plants that can be used for propagation.

6.2.1 Chemotherapy

In the introductory remarks to this chapter, attention was drawn to the difficulties of using chemotherapy to control viruses as compared to the chemical control of fungi. In spite of the problems, some attempts at chemical control of viruses have been made, often with moderate success. Commoner and Mercer (1951) found that the nucleic acid base analogue thiouracil inhibited the multiplication of TMV in detached leaves. Unfortunately, application of this chemical to whole plants seriously affected plant growth by interfering with cellular nucleic acid metabolism. For similar reasons other nucleic acid base analogues, although having antiviral activity, have proved impracticable for chemical control of plant viruses. The antimetabolic nucleoside analogue virazole (ribavirin) appears to block virus replication, and can eradicate virus from plant tissue. Shepard (1977) reported that virazole eradicated PVX in cultured tobacco meristems, whilst concentration of alfalfa mosaic and cucumber mosaic viruses are reduced in tissue cultures (Simpkins et al., 1981). In a slightly different approach Cassels and Long (1980), found that PVY and cucumber mosaic virus could be controlled in adventitious shoots derived from excised petioles of diseased tobacco plants. Whether virazole can be used to successfully control viruses in anything other than experimental and tissue culture systems, remains to be seen.

Numerous other chemicals have been tested against viruses, including the dyes malachite green and methylene blue, as well as nicotinic acid, amino acids, indoleacetic acid (IAA), 2:4 dichlorophenoxyacetic acid (2:4 D), hydroquinone, and various mineral salts (Carter, 1973). Milk, and various plant extracts, also influence virus multiplication. Plant growth substances, such as cytokinins and auxins, may reduce virus concentrations on plant meristems but do not eradicate virus (Walkley, 1980).

In studies involving TMV infection of tobaccos (*Nicotiana glutinosa* and *N. tabacum*) and beet western yellows virus infection of lettuce, it has been shown that the chemical carbendazim reduces virus-induced yellowing, but has little effect on virus numbers or on local lesion production (Tomlinson *et al.*, 1976). Carbendazim is the fungitoxic principal in the fungicide benomyl. The use of such compounds against virus may be more harmful than useful since viruses in the symptomless hosts are dangerous reservoirs of disease agents for other plants.

An alternative approach to chemical control of viruses might be possible if plants, like animals, produce natural defence mechanisms against viral invasion. Two types of antiviral activity have been described for plants. Firstly, plant extracts reduce virus activity when they are inoculated together with virus into plants in experimental situations; whether this is equally true when plants are naturally infected with virus by vectors is not known. The active inhibitory ingredient of the extracts has been variously identified as carbohydrate, amino acid, tannin, glycoprotein and protein. The best documented are those from pokeweed (*Phytolacca americana*) and from carnation (*Dianthus caryophyllus*), both of which are glycoproteins and inhibit protein synthesis by binding to ribosomes (Stirpe *et al.*, 1981). Such binding occurs only with ribosomes from plants different to the source of glycoprotein. Inhibition of virus replication may result from inhibition of the production of viral proteins or of the enzymes necessary for virus replication.

The second type of antiviral activity is the induction of antiviral factors or compounds (AVFs), following the inoculation or spraying of plants with either virus, chemicals or plant extracts. Such AVFs may be proteins or phosphoglycoproteins. Kassanis *et al.* (1974) found that induced resistance of leaves challenged with a second infection was correlated with the appearance of those proteins not present in healthy leaves. Similar proteins could be induced, and were associated with resistance, when leaves were treated with polyacrylic, benzoic, salicylic and acetylsalicyclic acids. Similar induction of AVFs was found when French bean and tobacco leaves were treated with *Gypsophila paniculata* extracts (Barakat and Stevens, 1981). It has been suggested that AVF from plants may resemble interferon. The AVF from *Datura* resembles animal interferon in terms of molecular weight and ability to prevent virus replication (Loebenstein *et al.*, 1966). This is an exciting area of study where more work should be undertaken. The fact that chemicals can be used to induce resistance means that a direct chemical method of controlling viruses based on the plant's own defence mechanism may be developed. Alternatively, antiviral factors might be extracted from plants and then applied to crops for protective purposes. Plant breeders may be able to produce varieties high in natural virus inhibitors.

6.2.2 Thermotherapy

The exposure of virus-infected plants to high temperatures for a period of several days often yields virus-free plants. We have seen earlier how virus can be eliminated from seeds by heat treatment. One of the earliest records of the successful use of heat therapy is that of Kassanis (1949) who showed that potato leaf roll virus was eliminated from potato tubers by heating them to 37°C for 25 days. Subsequently, many plants were heat-treated, and by 1965 Hollings was able to list about 90 plant diseases successfully treated by heat therapy. Plants such as raspberries, strawberries, apple, citrus and a number of ornamentals have been freed of virus by heat treatment. In general, rod-shaped or filamentous viruses are less susceptible to elimination by heat than spherical viruses. Dormant tissue such as bud-wood, or dormant bulb and tubers, can be treated by immersion in hot water. Growing tissues are best freed of virus by prolonged hot air treatment. Temperatures of 35–40°C appear optimal. These temperatures are not related to the thermal inactivation point of the viruses (section 7.3.2).

Alfalfa mosaic (AMV), potato leaf roll (PLRV) and tomato black ring (TBRV) viruses can all be eliminated from potato tubers by hot air treatment at 37°C for 3–6 weeks (Kaiser, 1980). Similar treatment for up to ten weeks did not eradicate PVY, however, and hot water treatment of tubers at 50°C for 25–180 minutes or 52.5°C for 15–90 minutes did not free tubers of AMV, PLRV or TBRV. In Kenya, tubers freed of virus by hot air treatment have been distributed in several potato improvement pro-grammes in East Africa.

Ornamentals such as carnations and pelargoniums (geraniums) can be freed of viruses by thermotherapy, and in the case of chrysanthemums, aspermy virus and chrysanthemum viruses B, D and E can be eliminated from plants by treatment at 38°C for 1–3 months.

Very often thermotherapy is used in combination with meristem culture techniques.

6.2.3 Meristem culture (Fig. 6.1)

Observations by Limasset and other workers in the late 1940s established that plant apical meristems could be considered as free of invading viruses.

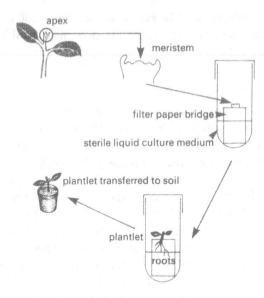

Figure 6.1   Diagram to show the production of plants by meristem culture.

By excision of the apex under aseptic conditions it proved possible to cultivate whole plants free of virus. Morel and Martin first used this meristem-tip culture method to produce virus-free dahlias in 1952. The first economically important application of this technique was probably that started in 1957 at Rothamsted Experimental Station to produce virus-free King Edward potatoes. King Edwards contained potato virus M in particular, and although the viral symptoms were hardly visible, the production of virus-free plants produced a 10% increase in yield, which it was calculated represented an annual benefit to UK agriculture of £2 million in 1965.

Virus-free plants have subsequently been produced by meristem culture from a wide range of species; Walkley (1980) lists over 40 species freed of virus. The technical details vary from worker to worker but the method usually involves the following stages:

(1) Careful selection of parent material and identification of possible virus infection.
(2) Thermotherapy of parent plant.
(3) Excision of meristem-tip (apex).
(4) Culture of apex on suitable medium to produce plantlets.
(5) Thermotherapy or chemotherapy of culture (e.g. virazole in medium)
(6) Plantlets transferred to soil.
(7) Maintenance of virus-free plants.

During stages (6) and (7), plants are carefully indexed to establish that they are virus-free.

The production of virus-free plants by meristem culture provides *nuclear stocks* of plants from which further virus-free plants can be obtained.

6.2.4 Certification schemes and nuclear stock propagation

There are systems of rules, regulations and check inspections which are used by government plant health authorities to maintain standards for the production of disease-free propagation material. Such systems of rules are called *Crop Certification Schemes*. Botanical purity and crop vigour as well as crop health are qualities controlled through certification schemes (Ebbels, 1979). Plants grown for propagation are subjected to growing-season inspections, where checks are made to establish that plants are in fact true to type and do not contain large numbers of mutants or consist of a mixture of cultivars. Plant vigour is examined and good crop husbandry reported upon. Good crop health is the priority however, and plants are examined and tested (indexed) rigorously for pathogens. Plants reaching the required standards are given certificates of health and are recommended for propagation purposes. The best known scheme in the UK, for example, is that governing seed potato production (Brenchley and Wilcox, 1979). Details of the health conditions required for potato certification are given by government plant health authorities.

With many vegetatively-propagated ornamentals, bulbs and fruits, virus-free stock plants (nuclear stock) are produced for distribution to commercial growers. In the USA and some European countries, indexing of plants and production of nuclear stock is undertaken by private industry working together with government-financed plant pathology departments. In America, for example, the State Department of Agriculture in Pennsylvania carries out test procedures for detecting viruses in pelargoniums, and in New Jersey there is a well-established certification scheme for the production of orchids free of cymbidium mosaic and TMV. In Great Britain, growers of ornamentals have formed a Nuclear Stock Association which works with government research establishments in producing virus-free stocks of chrysanthemums, carnations, pelargoniums, and bulbs. Similar nuclear stock associations operate for strawberries, hops, rhubarb and fruit trees. Some schemes are international in character, for example the Interregional Research Project (IR-2) initiated in Washington State University, is designed to produce virus-free cultivars and clones of deciduous trees and to distribute propagating material for research to industry. In 1980, propagation material had been distributed to 40 American states to five Canadian provinces and to 40 other countries via the United States Department of Agriculture and the Food and Agricultural Organization of the United Nations (Fridlund, 1980).

6.3   Control of vectors

Virus diseases can be controlled to some extent by controlling the vectors of disease. In the case of insect vectors and aphids in particular, control may take the form of (*a*) avoidance, and protection of plants from vectors; or (*b*) elimination of vectors.

6.3.1   *Avoidance and protection of plants from vectors*

This comes about through particular agricultural and horticultural plant culture and husbandry techniques, a number of which are outlined below.

Crop isolation. If crops are isolated by growing them some distance from other similar crops, then the chance of invasion by virus-carrying vectors is reduced. This is the reason why potato plants grown for seed certification schemes should be no nearer than 50 m from potato not in the certification scheme. Another example is that of young sugar beet plants, raised in areas isolated from beet and mangold crops to control yellows viruses.

Virus-free potato seed production is based on avoidance of vector aphids. In the UK, the best seed areas are in Scotland, where the low incidence of potato leaf roll virus (PLRV) is due to the lateness of aphid infections rather than to complete freedom from the virus. Young plants showing symptoms can be removed (roguing) before aphids enter the crop, and the low incidence of disease throughout Scotland means that the crop is generally very healthy; in fact, there were no serious epidemics of aphid-borne viruses in Scotland between 1945 and 1973. By roguing and certification procedures, negligible levels of PLRV were recorded in seed potato crops until 1970. By 1973 the levels of PLRV rose to 50% of crop and this was related to early migration of aphids following the mild winters between 1970/71 and 1974/75 (Woodford, 1977). We will return to this problem later.

Crops grown under glass are to some extent isolated or protected from outside sources of virus and of vector. Additional protection is achieved by installing double doors to glasshouses, by insect-proof netting over vents, and in some extreme cases by pressurization of houses to oppose insect entry. Within the glasshouse, however, overlapping of crops may result in virus spread between plants of different or the same type.

Barrier crops. Surrounding a crop with some other plant species may influence disease incidence. Barley, for example, grown around, or in rows between sugar beet seedlings, lowers the incidence of beet yellows. Similarly, the spread of insect-transmitted viruses to low-growing ornamental shrubs can be prevented by planting the shrubs between rows of rye. Vectors remain on the barrier plants and therefore the incidence of disease within the crop is lowered.

Reflective surfaces. Aphids are attracted by some colours and repelled by others. When the green peach aphid (*Myzus persicae*) is flying 2–3 ft above vegetation, it is attracted by yellow and, to a lesser extent, orange or green surfaces. Blue, violet, black, grey and white are ignored. The response to colour varies, however, between aphid species (Moericke, 1954). Whiteflies are also attracted to yellow surfaces. In other experiments, aluminium trays around yellow trays repelled aphids, possibly due to the high reflection of the metal (Kring, 1954). This feature has been used to repel insects from crop plants such as gladioli, lilies, squash plants, lettuce and cucumber, but proved unsuccessful with chrysanthemums (Smith and Webb, 1969).

Plant distribution and density. Viruses spread rather more efficiently along rows of plants than between rows, since the distance between rows is greater than that between plants in a row. Contact between plants may result in greater virus spread, as with cacao swollen-shoot virus where mealy-bugs crawl from plant to plant carrying the virus. With flying insects the incidence of disease is influenced by plant density.

When a given area contains a large number of plants, a smaller proportion will be infected by incoming populations of insects than when the area contains fewer plants. This situation may be more complex with some crops, and factors other than plant density may be involved. It has been found in ground nut, for example, that close spacing of plants gives a continuous plant cover which may inhibit aphid vectors from landing on the crop (A'Brook, 1973).

Field size. Insects entering crops often congregate on the margins of fields, which means that in large fields there is an internal area free of vectors and of virus. Thus, in small fields a larger proportion of the crop will be infected than in large fields.

Planting and harvesting dates. Manipulation of planting and harvesting dates offers some degree of virus control. Losses of cereals by barley yellow dwarf virus are highest if sowing is early (late September or the first week of October). Seed sown in mid-October escapes serious infection. Early-sown cereal is subject to infestation by migratory aphids. On the other hand, early-sown broad beans have lower incidence of pea mosaic virus than late-sown plants, because the older plants are more resistant to infection and may not be as attractive to aphids as younger plants.

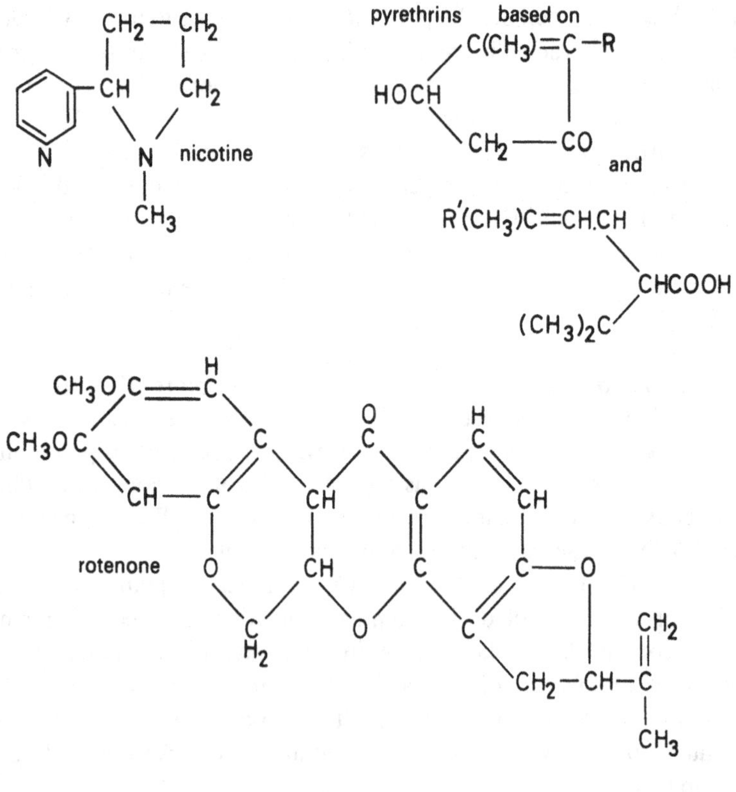

Figure 6.2 Formulae of some naturally-produced insecticides. Nicotine ($C_{10}H_{14}N_2$) is obtained from tobacco plants, particularly *Nicotiana tabacum* and *N. rustica*. Pyrethrins are obtained from *Pyrethrum cineraefolium*. Rotenone ($C_{23}H_{22}O_6$) is obtained from species of *Derris* and *Lonchocarpus*.

6.4 Elimination of vectors

Control of viruses by methods aimed at avoidance of, or protection from, vectors may not always be convenient or adequate. The elimination of vectors by direct chemical control may therefore be necessary. The effectiveness of such methods of control depends on a number of factors including type of vectors, type of transmission, and timing of application of chemicals.

6.4.1 *Control of insects by insecticides*

Insecticides can be divided into *natural products* and those that are *synthesized*. Natural products (Fig. 6.2) include nicotine derived from tobacco, pyrethrum derived from *Chrysanthemum cineraine-folium*, and rotenone-based compounds derived from the legume *Derris elliptica*. These compounds are used mainly to control insects on glasshouse crops,

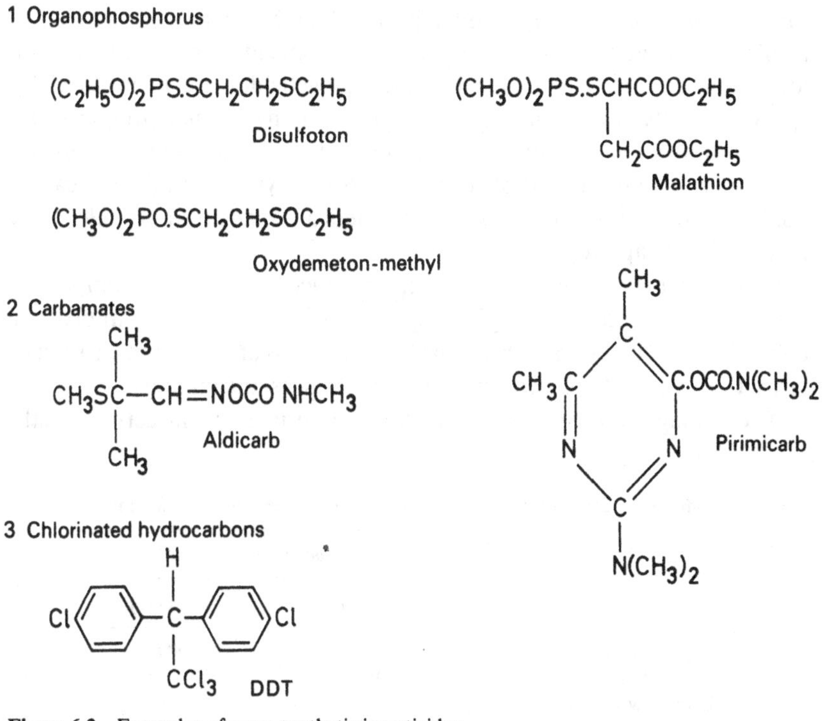

1 Organophosphorus

$(C_2H_5O)_2PS.SCH_2CH_2SC_2H_5$

Disulfoton

$(CH_3O)_2PS.SCHCOOC_2H_5$
$|$
$CH_2COOC_2H_5$

Malathion

$(CH_3O)_2PO.SCH_2CH_2SOC_2H_5$

Oxydemeton-methyl

2 Carbamates

CH_3
$|$
$CH_3SC-CH=NOCO\,NHCH_3$
$|$
CH_3

Aldicarb

Pirimicarb

3 Chlorinated hydrocarbons

DDT

Figure 6.3 Formulae of some synthetic insecticides.

although nicotine is used to control aphids on fruit trees. Because they are natural products it must not be assumed that they are less harmful to humans than synthetic pesticides.

Synthetic pesticides (Fig. 6.3) are largely based on three groups of compound.

1. *Chlorinated hydrocarbons.* Of these, DDT is the best known, but this and many related compounds have been withdrawn in the UK, Europe and USA because of their tendency to accumulate in tissue and then pass through food chains to birds and other animals, causing death. Dichloropropene and dichloropropane are compounds used as nematicides.

2. *Organophosphorus compounds.* These constitute the largest group of pesticides of value in virus vector control. They may be of short persistence, or they may be systemic in that they are transported about the plant.

3. *Carbamates.* These make up the third group of insecticides. Some, such as ethiofencarb and thiofanox, are systemic, others, such as aldicarb and pirimicarb, are not.

Pesticides may be applied in several ways, depending on their chemical formulation, the crop area and type of plant being treated, as well as availability of labour and equipment. Spray application may be by high or low volume spray (see Table 6.1). Some pesticides are formulated for spraying at ultra-low volume (ULV) using equipment which produces very small droplets (60–110 micron diameter range). In glasshouses, aerosols and fogs can be used to apply chemicals. Some systemic pesticides can be applied to the soil as drenches, whilst others are made into granules to be mixed with, or applied to, the soil.

Pesticides may damage plants if applied incorrectly. Some compounds are phytotoxic on certain plant species—malathion, for example, will control insects on chrysanthemums but is toxic to antirrhinums, *Crassula*, ferns, *Gerbera*, orchids, petunias, *Pila*, sweet peas and zinnias.

The damage to crops by non-vector or non-infective insects is usually

Table 6.1 Types of spraying technique used in insecticide application.

| | Spray applications (litre/ha) | |
	Bushes and trees	Ground crops
High volume	> 1120	670
Low volume	225–560	55–250
Ultra-low volume	< 225	< 55

proportional to the insect population and to the duration of feeding. The situation is more complex when insects are vectors of disease. If virus sources are within the crop, then the spread of virus will be determined by the intra-crop movement of the vector, the biology of the vector and the type of virus. With aphid vectors for example, infestations may be by winged forms (alatae) entering the crop from outside. Wingless (apterous) progeny tend to intensify disease within the crop. However, if winged forms are produced, spread within and beyond the initial areas of disease may be extensive.

If the virus is non-persistent, it can be acquired by aphids in a short time, possibly from epidermal cells, and can then be rapidly transmitted. To prevent the spread of such viruses, an insecticide that kills incoming winged aphids quickly is required. Aphicides are often effective in decreasing spread by wingless forms within crops but are not very effective in preventing infections introduced by alatae from outside sources. Contact insecticides seldom achieve sufficient cover to kill all aphids, and those that survive are able to transmit virus and also multiply and replace killed insects.

With persistent viruses, transmission is delayed whilst the virus circulates in the insect. If insects feed on plants treated with insecticide during the latent period, they may die before transmission takes place. Systemic insecticides can be used effectively, although chemicals that reside in plants for a long period are best avoided in food crops, except where plants are being grown for propagation purposes only.

Beet yellowing viruses and potato viruses are controlled reasonably well by insecticide. Careful monitoring of the crops for signs of vectors is required, however, to be successful. For instance, in the case of potato grown in Scotland, plants are examined for aphids by tapping the foliage for a few seconds over two white-painted rimmed boards held closely against the haulms. If the plants are too large, then the plants are counted on 50 to 100 haulms, or when the leaves touch within rows the top, middle and lower leaf is examined. Results are expressed as aphids per 100 plants or per 100 leaves for the three-leaf method. When five or more adult *Myzus persicae* or 10 or more *Macrosiphon euphorbiae* are recorded per 100 complete potato plants, then growers are advised to spray. Warnings are then issued in the local press and on radio and television, or cards are sent to growers, merchants and to spray contractors. In 1975 41 % of the main crop was sprayed and in 1976 80 % was sprayed, of which 15 % was also treated with granular insecticide at planting. This increased spraying resulted from the increased aphid population due to the milder winters of

1971, 1972 and 1974. Dates of spraying vary from year to year and from area to area depending largely on climatic conditions.

Insecticide application is not always effective, which may be due to mistiming in spite of spraying warnings, or inefficient spraying. Time of spraying is also critical for other crops. For example, autumn-sown cereals are at considerable risk from barley yellow dwarf virus (BYDV). Plants emerging from seed drilled in September and the first week of October are at high risk from autumn-migrating aphids, and spraying insecticide in the first half of November is most effective in controlling BYDV.

Care must be taken in the application of insecticide, not only because of ecological considerations, but also because indiscriminate use leads to the development of insecticide-resistant strains of aphids. This may be one of the reasons for increased incidence of virus diseases in potato crops in recent years.

Chemical control of insect vectors other than aphids has also been tried with varying success. Whiteflies, for example, can be destroyed in the field, and mung bean yellow mosaic virus so controlled, by applications of insecticide in mineral oil. Adult insects are easier to kill than larval forms, and younger, actively-growing leaves are more susceptible to infection than older ones. Vector control should be carried out frequently, particularly in the early phases of crop growth.

Glasshouse control of pests is to some extent easier, in that vectors can be detected more readily and treatment can be more intensive. Vector multiplication may, however, be more rapid in glasshouse conditions.

6.4.2   Control of insects by oil sprays

Aphids carrying non-persistent or semi-persistent viruses often fail to infect plants sprayed with water–oil emulsions (Bradley, 1963). These early observations are well authenticated, although the mechanism of inhibition of transmission by oil is not understood. It is known that oil films on leaves affect virus acquisition more markedly than they do inoculation; oil itself however appears to have no direct effect on viruses.

Application of mineral oil sprays is used commercially with success in the USA to reduce field spread of viruses in lilies. Aphid-borne tulip breaking virus (TBV), lily symptomless virus and cucumber mosaic virus (CMV) can spread quickly because lilies grow rapidly at the time flying aphids migrate. Spread of the persistent TBV and the non-persistent CMV between lilies has been reduced in the Netherlands by oil sprays, and the spread of virus in potatoes, peppers, cucumbers and tomatoes has been controlled in the USA. Oil sprays may also be of value in controlling whitefly-transmitted viruses (Nene, 1973). One feature limiting use of oil sprays is their phytotoxicity, and more information is required on formulation, methods and rates of application before more effective use of oil sprays can be achieved.

6.4.3 Control of mites, fungal and nematode vectors

Mites that transmit viruses can be controlled to some extent by application of organophosphorus pesticides such as demeton-S-methyl, dicofol, tetra-difon or oxydemeton-methyl.

Fungal and nematode vectors can often be treated together, using soil sterilization techniques. In glasshouses soil may be heat sterilized, but where larger volumes of soil are involved or it is more convenient, chemical sterilization may be appropriate. *Polymyxia graminis*, the fungal vector of wheat mosaic virus, can be destroyed with carbon disulphide or formaldehyde. Larger areas can be treated with methyl bromide, another general soil sterilant. The chlorinated hydrocarbons dichloropropene and dichloropane, as a mixture (D-D), or dichloropicrin, are compounds that kill fungi and also nematodes. Nematicides are the most practical way of controlling many nepo- and tobraviruses. Nematodes, being plant ecto-parasites, are readily exposed to chemicals applied to the soil, although different genera and species, as well as different stages in the life cycle, vary in their susceptibility to nematicides—for example, larvae of *Xiphenema index* seem to be less sensitive than adults to methyl bromide treatment.

Nematicides may be applied by fumigant or non-fumigant methods. Fumigants, such as methyl bromide and D-D, move through soils in the gaseous phase, and are therefore influenced by soil moisture content and soil particle size, so that clay soils are difficult to treat. Viruliferous nematodes have been found as deep as 3.6 m in Californian soils, although in general the largest populations are at a depth of 0.5 m. Fumigant can reach to depths of about 2.5 m, so that nematicide treatments are generally but not always successful in controlling vectors. High concentrations of these nematicides are phytotoxic and should be applied to soils only before planting.

Non-fumigant nematicides move through soil in water and can therefore penetrate deep soils. On the other hand, abundant rainfall may wash them to deep layers of soil beyond the zone of nematodes. Non-fumigant nematicides, such as aldicarb and oxamyl, are non-phytotoxic and nematostatic rather than nematicidal in action; they may simply delay feeding and egg-laying and reduce motility. These compounds can be applied in granular form and have the added advantage of acting as insecticides.

In the UK, arabis mosaic virus on hops, and tobacco rattle virus in potato crops, are controlled using D-D. Alphy (1978) showed that oxamyl does not decrease trichodrid nematode numbers, but does lower the incidence of nematode-transmitted tobacco rattle virus causing spraing of potatoes. D-D and chloropicrin are recommended for the control of *Xiphenema diversicaudatum* which is the vector of arabis mosaic and strawberry latent ringspot viruses in strawberries. In California, France and Germany, good control of grape fan leaf virus has been achieved for many years using D-D (Lamberti, 1981).

6.5   Other means of control

6.5.1   *Breeding for resistance*

A form of virus control of considerable interest and which may offer the best long-term answer to disease problems is the production of resistant plants. What is meant by 'resistant'?

Some plants may not be infected by a particular virus because the virus fails to replicate. Such plants are non-hosts and are *immune* to that virus. Plants which allow some degree of replication, however small, are *susceptible* and are host plants. Some hosts allow replication of virus with the production of conspicuous symptoms and distinct changes in host growth and yield; these are *sensitive* or *non-resistant* plants. *Resistant* plants are less liable to become infected, but if invaded by virus are able to minimize virus effects, so that symptoms are less severe than shown by *sensitive* plants. However, when virus replication occurs without external symptom production or change in growth and yield, the host is *tolerant*. Resistance may take the form of hypersensitivity where virus replication produces localized death of tissue, shown by local lesions, which stops systemic spread of the invading pathogen. In other cases resistance may be to *infection*, involving some form of inhibition of virus uptake and entry, or resistance can be to *replication*, which might well involve the activity of antiviral compounds or factors (AVF). Resistance can also be to the *vector* of the virus.

Resistance shown under laboratory or glasshouse conditions, where deliberate inoculations with cultures of the pathogen are involved, should be distinguished from *field resistance* where natural infection of plants is more dependent on environmental factors and also the biological status of both pathogen and vector.

The aim of plant breeders, then, is to develop varieties with some degree of resistance either to the virus or to its vectors. Plants completely immune would be the ideal achievement of the breeder.

Russell (1978) draws attention to the fact that plants are in fact immune to most viruses, and that susceptibility is the exception. As an example, he quotes sugar beet and potatoes as being immune to barley yellow dwarf virus (BYDV). This fact offers little hope to the plant breeder, unless immunity in beet or potato can be transferred by hybridizing cells of these with barley cells to produce BYDV-immune barley. Alternatively, if some wild relative of barley is already immune to BYDV, then this might be successfully interbred with susceptible cultivated forms to produce immune plants. This latter approach may be the most practicable, although advances in protoplast and tissue culture techniques may mean that the former idea is feasible in the future.

If breeders are able to produce plants showing some degree of resistance to virus or vector, then disease losses can be reduced. The production of tolerant varieties carries with it some element of risk, since these varieties may act as undetected reservoirs of virus, potentially able to infect other crops. Virus in tolerant plants may mutate to more virulent forms, or invasion by a second virus may produce more severe symptoms than either virus alone. Plants showing hypersensitive reactions are useful, since virus replication is limited and infected plants easily recognizable. Sensitive hosts that die quickly as a result of disease may also be valuable, as the virus is eliminated with the demise of the host.

The production of resistant varieties begins with selecting individual plants that show some hindrance of virus replication by mollified symptoms. Resistance may be located either in wild species, by chance observation in cultivated plants, or by testing large numbers of plants by deliberate exposure to virus or vector. Potentially useful plants are further tested for virus content, vector frequency and yield. Care must be taken to see that desirable characters in the plant are not lost in gaining disease resistance. Breeding for resistance is a long and often complex exercise, but the end results are often worth the effort, particularly since it has been found that in general resistance to viruses is more durable than, say, gene-resistance to airborne fungi (Harrison, 1981).

Successful use of virus resistance has been reported for a wide variety of crop plants. One of the earliest successes was with sugar cane, where varieties resistant to mosaic have been available for over 50 years

(Brandes, 1925). The development of sugar beet varieties resistant to curly top virus saved the sugar beet industry from extinction in large areas of the western USA. Resistant varieties were first isolated in the mid-1920s and further highly resistant varieties were obtained from these by 1931. Other major crops, such as beans, cassava and sweet potato, have also been subject to breeding for virus disease resistance. The necrotic local lesion reaction of *Nicotiana glutinosa* has been bred into some commercial tobacco varieties for TMV control purposes. In potato, extreme resistance to viruses A, S, X and Y and field resistance to tobacco rattle and to potato leaf roll viruses are now available in commercial cultivars (Harrison, 1980).

There are three major seed-borne viruses of French beans, namely bean common mosaic, bean yellow mosaic and tobacco necrosis virus. Cultivars such as Processor and Masterpiece are resistant or tolerant to these viruses, although no cultivar is completely immune or symptomless when infected (North, 1979).

Considerable work has been done on resistance to tomato mosaic virus. Tomato mosaic virus is very closely related to tobacco mosaic virus, but can be differentiated on host responses, serological reaction and coat protein amino acid composition. (Confusingly, tomato mosaic is often abbreviated to TMV—to avoid this confusion I shall abbreviate tomato mosaic to ToMV.) Isogenic lines of Cruigella tomato contain genes Tm-1, Tm-2 and $Tm2^2$ for resistance to five different strains of ToMV. Strain 0 is unable to overcome any of the resistance genes; strain 1 can overcome gene Tm-1 so that plants with this gene show symptoms. Strains 2, 2^2 and 1.2 overcome genes Tm-2, $Tm2^2$ and both Tm-1 and $Tm2^2$ respectively. Only virus strains 0 and 1 are found commonly. Early resistant varieties such as Virocross and Supercross contained Tm-1, and after only one season resistance was broken, as a change occurred in the frequency of virus strains. When Tm-1 resistant varieties were introduced there was almost 100% strain 0 virus, whereas two years later, 39% of the population was strain 1, capable of overcoming resistance of Tm-1 plants. Where cross-protection (see later) with ToMV M11–16 (a strain derived from strain 1) had been used, the incidence of strain 1, and therefore of resistance breaking, increased to 94%. Best resistance can be achieved using all three Tm genes. Modern commercially-grown varieties, such as Sarine, Sonato, Sonatine, Nemato, E4884 (Dawn) and the 'J' cultivars from the Glass-house Crop Research Institute, are ToMV resistant, although in Guernsey the non-resistant variety Grenadier was the main variety grown until the 1981 season, when E4884 took its place.

The breeding of plants resistant to virus vectors has been attempted for several crop plants. Resistance of cotton plants to leaf curl is partly due to resistance to the whitefly vector *Bemisia* spp. Several virus diseases of rice are transmitted by leaf hoppers, particularly *Nephotettix virescens*. Some rice varieties such as Pankhari 203, A5D7 and IR8 are resistant to *N. virescens*. 90% of nymphs died on these plants and those surviving developed very slowly.

Resistance to aphid vectors has been reported from sugar beet, cabbage, potato and raspberries. With sugar beet, resistance to *Myzus persicae* and *Aphis fabae* may be due to preference—that is, choice of plant upon which to settle—or to multiplication of aphids (antibiosis). Non-preference (resistance to settling) of *M. persicae* may not be associated with non-preference in *A. fabae*. Resistance may vary between wingless and winged forms (Russell, 1978).

Resistance of potatoes, *Solanum tuberosum*, to virus-carrying aphids is found in several wild species of *Solanum*. Resistance in *S. taryense* and *S. berthantii* is associated with the presence of four-lobed glandular hairs on the leaves; some have longer hairs with glandular tips. The presence of these hairs is controlled by a single dominant gene. Plants with sticky-tipped hairs had 60% fewer *M. persicae* than plants with non-sticky tipped hairs (Gibson, 1978). Resistance may be partly due to non-preference, together with the fact that aphids become immobilized by the sticky fluid exuded by the hairs and starve to death. If the glandular hair character can be bred into cultivated varieties this may produce effective resistance against virus-carrying aphids.

Some degree of resistance to most of the damaging raspberry viruses is obtained in Europe by resistance to aphids (*Amphorophora idaei*) of which four strains are known. Major genes are derived from *Rubus idaeus*, particularly Baumforth and Chief cultivars. Gene A1 confers resistance to aphid strains 1 and 4, gene A2 resists strain 2. Genes A8 and A9 derived from *R. idaeus strigosus*, and genes A10 and A11 from *R. occidentalis* and *R. coreamus* respectively, confer resistance to all four aphid strains. Many commercially available raspberry varieties are now resistant to *A. idaei*.

6.5.2 Cross-protection

Breeding resistance to virus or vectors is complex because many other characteristics of the plant must be considered in the breeding programme. As mentioned earlier, the non-virus-resistant tomato variety 'Grenadier' has been the most widely grown in Guernsey until very recently. Why has it not been replaced earlier? The answer lies in the fact that other characteristics of Grenadier such as reliable growth, fruiting capacity and fruit character have meant that, from a commercial point of view, this variety is easier to grow and more cost-effective than previously available virus-resistant cultivars. The problem of virus control in commercially acceptable virus-susceptible varieties has been achieved by the deliberate inoculation of plants with a mild strain of ToMV. Rast (1972) produced a nitrous acid mutant strain of ToMV, strain M11–16, which is symptomless in tomato and can protect plants against infection by more damaging strains. This is called *cross-protection*. One of the first commercial uses of this technique was made in the Isle of Wight in 1965 when early deliberate inoculation of tomatoes with ToMV reduced from 30% to 3% the quantity of unsaleable fruit spoilt by late virus infections (Broadbent, 1976). These early control methods used the severe strain since mild strains were not available.

In Guernsey, mild strains of ToMV are sold for use by commercial growers. The mild strain was originally obtained from Holland, but since Dutch growers now rely on virus-resistant plants, Guernsey produces its own M11–16 virus. Cross-protection has also been used in attempts to control viruses of other crops, including tobacco in America and cacao in Ghana.

6.5.3   Biological control of vectors

This is a popular and attractive alternative approach which is being actively researched. Nene (1973) explored the possible control of whitefly virus vectors by the parasitic fungus *Paecilomyces farinosus*. In Europe, whiteflies can be controlled by a parasitic wasp, *Encarsia formosa*, although this is difficult, particularly in early spring. The possibility of reducing infestations with the fungus *Verticillium locanii* early in the season, before introducing *E. formosa*, is being investigated (Kanagaratum *et al.*, 1979). This same fungus is also being tested for use against aphids, particularly *Aphis gossypii*, on chrysanthemums. Other fungi are also being tested against leaf hoppers and thrips (Burgess, 1981).

6.5.4   Integrated control

In designing or accepting measures to control virus diseases, care must be taken to judge carefully the expense and expertise required for any particular control method against crop losses. Vigilance must be exercised since there may be need to change the control strategy in crops as agricultural and horticultural practices change. A single approach to control is unwise. For example, crop hygiene should not be abandoned simply because resistant varieties are being used. Integrated control of disease by use of virus-free or virus-resistant varieties, plus good crop husbandry and sensible use of pesticides, will lead to higher crop yields with enhanced crop health and quality.

CHAPTER SEVEN

TECHNIQUES IN PLANT VIROLOGY

Let us imagine that we have been presented with diseased plant material and asked to identify the causative agent. If we also imagine that tests have shown that no fungal, bacterial or mycoplasmal pathogens are responsible for the symptoms, neither are they the result of nutrient imbalance or pest or spray damage, how then can we

(a) determine that a virus or viruses are involved, and

(b) identify the virus?

The following is an outline of some of the methods and techniques that can be employed.

The extent of any investigation will depend on the amount of plant material available. Initially, crude extracts can be

(a) examined in the electron microscope for virus particles;
(b) mechanically inoculated on to a range of plants to ascertain symptom production and host range;
(c) reacted with serum containing antibodies to known viruses, to establish serological relationships.

Attempts can be made to establish that only one virus type is involved and that it can be re-inoculated into the original host species to produce symptoms similar to those originally described. Crude extracts can also be used to study the *in-vitro* properties of the virus.

Following these tests, it may be possible, using a suitable host, to propagate the virus and make pure preparations. The virus may then be characterized and compared with known viruses. A cryptogram could also be written and the virus assigned to a particular virus group.

We shall consider these preliminary tests first. Crude extracts are prepared by grinding tissue in water or buffer, and centrifuging or filtering through muslin to remove large cell debris.

7.1 Electron microscopy

This will give a rapid idea of the type of particle involved, provided sufficient virions are present in the extract.

Crude extract or sap from the infected plant is placed on a copper grid covered by a support film of carbon and/or plastic, such as formvar. The virus particles are transparent to the electron beam, so they are mixed with a solution of heavy metal salts such as 2% potassium phosphotungstate (KPTA) or 1% uranyl acetate or similar material. Such 'stains' form an electron-dense background so that virus particles show up as clear structures on a dark field. This is called *negative staining*. The metal salts also penetrate the particles and enable sub-structural details to be observed. Some viruses are disrupted by treatment with heavy metal salts, but the particles can be stabilized by treatment with 10% formaldehyde for 20 minutes. Where rapid examination is required, or only minimal infected material is available, a freshly broken edge of the infected tissue can be drawn through a drop of heavy metal 'stain' on the grid. The excess stain is blotted gently from the grid and it is viewed in the electron microscope. Virus extracts mixed with stain can also be sprayed with an artist's airbrush or similar device on to grids, air dried and viewed in the electron microscope. If sufficient virus is present in the sample, particles may be found, so that it can be established whether isometric or elongated viruses are present.

7.1.1 Shadowing techniques

Shadowing (Fig. 7.1) consists essentially of spraying the object under study with heavy metal, such as gold or gold-palladium, from an angle in a vacuum chamber. The specimen is examined in the electron microscope and the shape of the object shows up as electron-dense 'shadow' on less dense background. This process is more difficult than negative staining and is used less frequently. Shadowing, however, forms the basis for studies of the length and form of nucleic acids removed from viruses.

7.1.2 Thin sections

Thin sectioning will be mentioned here as we are dealing with electron microscope techniques.

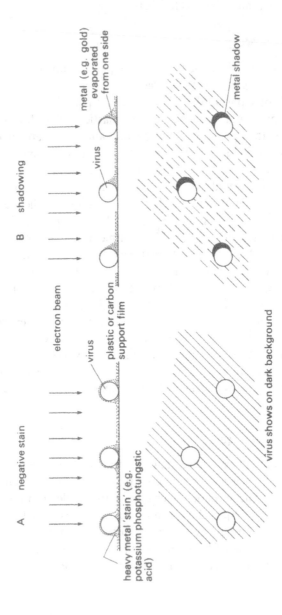

Figure 7.1 Electron microscopy of viruses. Preparations of PVX, (*A*) negatively stained, and (*B*) shadowed.

Briefly, thin sections of infected leaves can be examined for virus inclusions and for virus particles. The technique is time-consuming and involves fixation in glutaraldehyde, followed by treatment with osmium, dehydration, and embedding in a suitable resin. Sections of the resin block are cut with glass or diamond knives. Sections may be counter-stained, using lead citrate for example, and then viewed in the electron microscope. Preservation of crystalline arrays of virus may require glutaraldehyde-uranyl acetate mixtures or more dilute osmium tetroxide, than conventional electron microscope preparative techniques. Warmke and Christie (1967) recommend for example 0.02 to 0.08 % osmium tetroxide instead of the 2 % usually used.

Inclusions may be characteristic of some groups of viruses (Hamilton *et al.*, 1981). Some cytopathic effects, such as the presence of crystalline inclusions, can be observed by the light microscope and it is advisable to make preliminary comparison between healthy and infected plant material in this way before attempting electron microscopy. Techniques involving combined electron microscopy and serology are discussed later (section 7.6).

7.2 Host range

Crude extract, mixed with a small quantity of fine-grade carborundum (600 grit) or Kieselguhr to act as an abrasive, is rubbed on to the leaves of a range of possible host plants (see 3.1.1). Table 7.1 lists plants commonly employed in this test. These plants are examined for virus induced symptoms. Local lesions may appear in 2–3 days or longer; systemic symptoms may take several weeks to appear, depending on the host and environmental conditions.

Table 7.1 Plants commonly used to test virus host range.

Nicotiana tabacum v. Xanthi nc	*Phaseolus vulgaris* (French bean)
N. tabacum v. Samsun NN	*Vigna unguiculata* (cowpea)
N. tabacum White Burley	*Vicia faba* (broad bean)
N. glutinosa	*Cucumis sativus* (cucumber)
N. clevelandii	*Cucurbita pepo* (pumpkin)
Lycopersicon esculentum (tomato)	*Brassica campestris* (turnip, tender green
Solanum tuberosum (potato)	mustard)
Petunia hybrida	*B. pekinensis* (Chinese cabbage)
Gomphrena globosa	*Pisum sativum* (pea)
Chenopodium quinoa	*Triticum aestivum* (wheat)
C. amaranticolor	*Hordeum vulgare* (barley)

Although host range and symptom expression are not very precise indicators of virus identity, they are useful in discovering local lesion hosts for quantitative manipulations, and also for finding systemic hosts where virus may be propagated for subsequent purification.

Host range studies using mechanical means of transmission only are not altogether satisfactory, since some viruses are not easily mechanically transmitted. Other methods of transmission, such as passage by aphids, leafhopper and nematodes, should be considered. These means of transmission, and particularly those involving fungi, seed and pollen, may require specialized techniques and involve time-consuming experimentation. These studies are essential to the full description of a new virus, although the identity of a recognized virus may be established without a full description of the particular virus isolates/vector relationships.

If the virus can cause symptoms in other hosts, then this is helpful in determining whether the disease is the result of a single virus or a virus mixture. Electron microscopy of the original host and of test plants may help to establish this. If single lesions are cut from the local lesion host, crushed in a drop of buffer, then rubbed on to further host plants, it is possible to separate one virus from another. Individual virus types from lesions should also be inoculated back on to the original host species to determine whether the original symptoms can be obtained. Failure to produce typical symptoms with single isolates may indicate mixed infection by two or more viruses.

Test on the original host should also be carried out when virus has been purified. Koch (1880) formulated four postulates or rules regarding the identification of plant pathogenic bacteria; ideally these postulates should be satisfied in identifying any plant pathogen. These rules of proof involve:

(1) The constant association of the organisms with the disease.
(2) Isolation of the organism in pure culture.
(3) Reproduction of the disease using the cultured organism.
(4) Re-isolation of the organism from the diseased host and identification of it with the original inoculate.

It is not possible to produce plant viruses in a pure culture, and therefore, strictly speaking, Koch's postulates cannot be fulfilled. However, it is possible to work within the spirit of these rules and so establish the identity of the causative virus.

7.3   Studies *in vitro*

Crude extracts of virus can be characterized in respect of dilution end point, thermal inactivation point and longevity *in vitro*.

7.3.1   *Dilution end point* (DEP)

This is found by inoculating 10-fold dilutions of the crude extract on to local lesion host plants to find the highest dilution producing symptoms.

7.3.2   *Thermal inactivation point* (TIP)

Separate samples of crude extract are each heated separately to a temperature of 50, 60 ... 90 or 100°C for 10 minutes and then quickly cooled. Each sample is inoculated on to a local lesion host. The temperature treatment inactivating the virus so that no lesions are produced is the thermal inactivation point of the virus.

7.3.3   *Longevity* in vitro (LIV)

This is the length of time that crude virus extract remains infectious when kept at room temperature (20–22°C). The LIV of many viruses is measured in weeks. Preliminary tests are carried out on virus stored at room temperature after 3, 6, 18 or 24 hours for example; this is then extended to 2, 4 and 6 days, and finally weekly tests are carried out.

These characteristics (DEP, TIP and LIV) are not very precise, and their value is in doubt (Francki, 1980). Such tests may have some value, however, in judging conditions for purification or for separating mixtures or viruses.

7.4   Serology

When protein is injected into the bloodstream of a mammal, the animal reacts by producing *antibodies* which can bind to and inactivate the foreign protein. Plant viruses injected into rabbits will induce such a reaction since the virus coat is proteinaceous. The antibodies formed are immunoglobulins (Ig) and combine with special parts of the surface of the virus at some specific amino-acid sequence (the *antigenic site*), the virus protein being the *antigen*. Blood serum containing antibodies is called *antiserum*. Antibodies in this serum will bind with antigen to produce a precipitate and this is the basis of serological tests for viruses. The tests are highly specific since antibodies will only react with their specific viral antigen (Fig. 7.2).

Antiserum produced against a particular antigen (the virus) is called *homologous* antiserum with respect to that antigen. If this serum also reacts

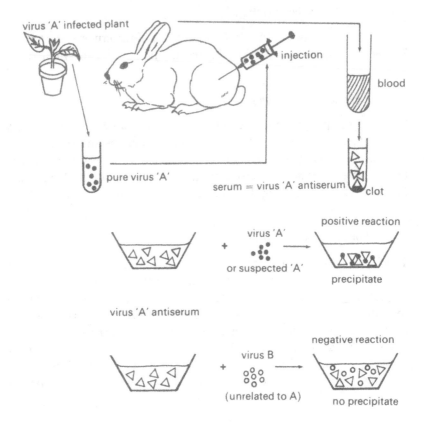

Figure 7.2   Antibody-antigen interaction.

with another antigen (another similar virus), the serum is then *heterologous* with respect to that virus. The fact that antibodies in the serum have reacted with another virus suggests that the viruses have some common antigenic sites, and are serologically related.

In performing these tests suitable controls must be set up. These include reacting sap from similar virus-free plants with the antiserum. Also included in the test should be *pre-immune* serum taken from the animal prior to immunization with the virus being tested, or *normal serum* taken from a similar animal not treated with virus.

Antiserum is produced by injecting rabbits either intravenously or intramuscularly with purified virus. Repeated intravenous injections are given at short intervals, for example 1 ml ($\simeq 2$ mg) of virus are given on three consecutive days, repeated twice after four days' rest (Ball *et al.*,

1974). Intramuscular injections are carried out by emulsifying the virus preparation with Freund's adjuvant, a mixture of mineral oil and surface-active agent, which increases the yield of antibody and may prolong its production. Injection schedules vary considerably from worker to worker depending on the virus being used. Generally, antibody production increases progressively and reaches a maximum 4–6 weeks after the initial injection.

Serum is obtained by taking blood samples from the rabbit. Intravenous bleeding, like injections, can be carried out via the marginal vein of the ear. The vein is dilated by warming or rubbing gently and then a small incision made and blood collected in a centrifuge tube. With kind and gentle handling the rabbits show no distress and suffer no more than a slight pricking sensation at the time of injection or bleeding.

The blood is allowed to clot at room temperature for 3–4 hours or at 37°C for 30 minutes. Storage in a refrigerator at 4°C overnight results in shrinkage of the clot, and the clear serum can be decanted and mixed with an equal volume of glycerin and stored at 4°C. If a number of sera are available then these can be used to test crude plant extracts for virus. Numerous tests have been devised; a few of the more important ones are described below.

7.4.1 Precipitin tests

When antigen and antibody are mixed they combine and form a precipitate. This precipitation or precipitin reaction is widely used in plant virology. The extent of the precipitate formed is dependent on a number of factors, e.g. salt concentrations, pH, temperature, and presence of interfering compounds, and the ratio of concentration of antibody and antigen is important as excess of either may prevent precipitation (Fig. 7.3). Most rapid precipitation occurs at a definite antigen/antibody ratio called the *optimal proportion ratio*. The greatest dilution of antigen (virus) to produce a visible precipitate is called the *antigen dilution end-point*. The greatest antiserum (antibody) dilution giving a visible precipitate is the *antiserum titre*. The titre of one reactant is usually measured when the other is in dilute solution, since excess would inhibit the precipitation.

There are a number of precipitin tests. Some are outlined below.

Precipitin tests in agar. In the *double diffusion* test, high grade agar is prepared in buffer pH 7.6 containing 0.85 % NaCl in a preservative such as 0.02 % sodium azide. The agar is poured into plastic petri dishes to a depth

Figure 7.3   Precipitation diagram of an antiserum produced by injecting a rabbit with turnip yellow mosaic virus. Data based on that of Matthews, 1957. Figures in diagram represent time in minutes for first visible precipitate during incubation at 50°C. Figures in **boldface** indicate optimum proportion ratio i.e. mixture giving most rapid precipitation.

	Virus dilution (initial concn. 5 mg/ml)									
	1/1	1/2	1/4	1/8	1/16	1/32	1/64	1/128	1/256	1/512
1/1	3	1	**0.7**	2.7	12	49	500	*inhibition by antiserum excess*		
1/2	17	11	3.5	**2.7**	11	23	97	530		
1/4	52	40	27	15	**7**	17	39	290	—	—
1/8	166	116	84	56	32	**22**	34	105	—	—
1/16	—	245	197	137	128	90	**52**	90	120	—
1/32	—	—	550	348	325	280	196	**130**	196	—
1/64	*inhibition by virus excess*				—	—	—	430	**400**	—
1/128					—	—	—	—	—	—

Antiserum dilution (row axis label)

of 2–3 mm, or on to glass slides to a depth of 1 mm. Wells are cut in the set agar, either with a special cutter or with a cork borer, in the pattern shown in Fig. 7.4.

Virus samples are placed in the outer wells and antiserum in the centre well. The plates are incubated in a moist atmosphere at constant temperature. Viruses diffuse in a ring from the outer wells, and antiserum from the centre well. If the virus has antigenic sites in common with the antiserum, then a precipitate will form in the agar and this becomes visible between the two wells. Unrelated viruses show no precipitation. Various patterns of precipitation lines can be obtained depending on the relationship between the antigen and antibody (Fig. 7.4).

Because elongated viruses diffuse rather poorly through agar, the techniques must be modified either to use lower concentration agar (0.5 %), or the virus must be disrupted to produce small, more diffusable fragments.

In the *radial diffusion* test, antiserum or antigen can be incorporated into agar. Wells are cut in the agar to contain the antigen, diffusion occurs and where the concentration of the reagents reaches the right proportions, a precipitate is formed. In a recent application of this technique to test cereals for barley stripe mosaic virus (BSMV), antiserum to BSMV (a rod-shaped virus) is incorporated into agar, together with a virus dissociating agent, in 90 mm petri dishes. Small samples of cereal leaves under test are

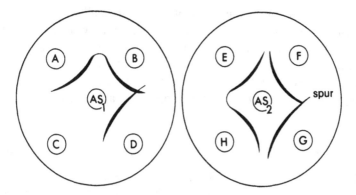

Figure 7.4 Double-diffusion test in agar. AS_1 and AS_2 = two different antisera (antibodies). $A, B...H$ = virus preparations (antigens). AS_1 reacts with A, B and D. AS_2 reacts with E, F, G and H. AS_1 has no antibodies reacting with C antigens. Precipitation lines A and B join so A and B combine with the same antibodies. E and H share the same antibodies in AS_2. Lines in B and D cross, showing that these wells contain two distinct antigens reacting with two distinct antibodies. Lines between F and G show partial fusion—spur formation—indicating that F and G share some antigens and differ in others.

inserted into the set agar about 1 mm apart in rows 2 mm apart; in this way about 400 samples can be tested. The plates are incubated for 24–36 hours and then viewed with a binocular microscope for the presence of a white precipitate. This test, which can also be used to assess cereal grains for virus, can detect 1 µg of BSMV per ml, approximately a 10-fold greater sensitivity than double-diffusion tests.

Precipitin tests in liquid media. In *tube precipitation,* a small volume (0.25–0.5 ml) of virus suspension is mixed with an equal volume of antiserum in small test tubes (0.7 cm diameter). Various dilutions of each reactant can be tested against each other. The samples can be incubated at 37°C in tubes only partially immersed in the water bath in order to cause mixing by convection. Some plant proteins precipitate at this temperature, so pre-treatment of sap for 10 minutes at 45–50°C followed by centrifugation clarifies the sap. Interaction of virus sample (antigen) and virus specific antiserum to produce a precipitate indicates a relationship between the unknown virus and the virus used in antiserum production.

In the *ring interface precipitin test,* antiserum is placed in a small tube and antigen is then layered on top. Diffusion occurs at the interface, and a disc of precipitation occurs where reactants have reached optimal proportions.

The *micro-precipitin test* is used commercially to test potato for PVX and other viruses. The technique involves placing small drops of antiserum on plastic or plastic-coated petri dishes, or in small wells in plastic trays (haemagglutination trays, Fig. 7.5). Plant extract is added to the antiserum and the drops viewed under a microscope to observe precipitation. Different antiserum–antigen ratios can be tested. Drops in petri dishes are usually covered with liquid paraffin to prevent evaporation.

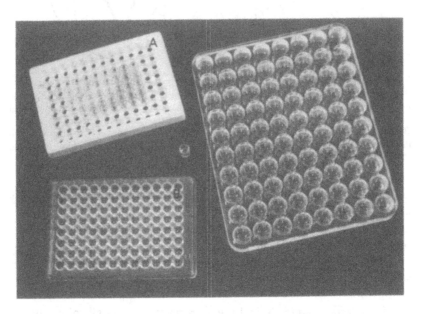

Figure 7.5 Microtitre and haemagglutination trays used for plant virus serology. (*A*) Individual microtitre wells in a polystyrene tray (one well removed). (*B*) Microtitre tray used for ELISA (see text). (*C*) Haemagglutination tray.

Chloroplast agglutination. These tests can be performed on crude sap extracts of suspected virus-infected plants. Plant material is ground to produce a sap suspension of chloroplasts and cell debris. Antiserum to the suspected virus is added and the mixture left for a few minutes. If viruses are present, the chloroplasts agglutinate because they are trapped in precipitate formed by interaction of virus and specific antibody.

This test tends to produce false positives due to non-specific reactions with components of the healthy plant sap. Although it is wasteful of antiserum, it is still used to index potato viruses such as PVX, PVS and PVM.

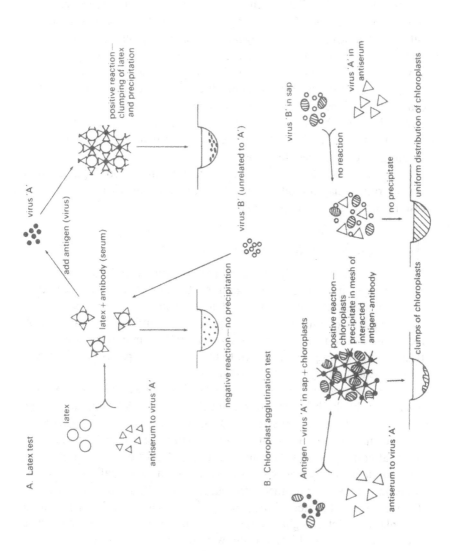

Figure 7.6 Latex and chloroplast agglutination tests.

Latex agglutination. In this test antibody is adsorbed onto polystyrene latex beads (0.81 µ diameter). The sensitized latex is then mixed with suspected antigen, and flocculation of the latex indicates antigen–antibody interaction (Fig. 7.6). Best results are obtained if the protein (globulin) of the antiserum is used. This is prepared by precipitation with ammonium sulphate. The advantages of using latex coated with globulin is that smaller volumes of antiserum are used, the sensitized latex can be stored, and sensitivity is increased over tube and microprecipitin tests. Low titre sera can be used in this test if the latex beads are first treated with protein A, a cell wall component of *Staphylococcus aureus.* Protein A binds to latex, then links to antibody molecules without affecting their ability to further link with antigen. The latex test is used extensively to test potatoes, barley seedlings and apple trees for viruses.

Care must be taken with the latex test since too high a virus concentration may inhibit flocculation. Inhibition or false positives may result from substances in plant sap.

7.4.2 Labelled antibody techniques

Antibodies may be labelled by conjugating them with radioactive compounds or ferritin (an iron-containing protein) or with a fluorescent dye. Such antibodies can then be used to detect the position of virus particles in tissues and to quantitatively measure antigen–antibody interactions. A recent innovation is the use of enzyme-linked immunosorbent assay (ELISA). This technique can be used to test plants for a particular virus, and is now used routinely to test plums for plum pox virus.

Several versions of the technique have been described, but the most usual is probably the direct double antibody sandwich technique. Briefly, the test is as follows. Antiserum against a particular virus of interest is treated with ammonium sulphate to precipitate the γ-globulin antibodies, which after dialysis are resuspended in buffer. A portion of the antibody is then conjugated to an enzyme such as phosphatase. Small plastic wells, either individually or in microtitre plates (Fig. 7.7), are used for the test. The wells are first coated with antibody, then washed, and the wells treated with plant extract thought to contain virus (antigen). The excess antigen is washed from the cells and each well treated with antibody–enzyme conjugate. The cells are washed free of unbound conjugate. Finally the wells are filled with *p*-nitrophenol phosphate substrate upon which the enzyme in the conjugate can work. The end product of the reaction, *p*-nitrophenol, is yellow in colour and so can be measured quantitatively.

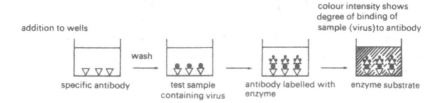

Figure 7.7 Enzyme linked immunosorbent assay (ELISA) tests. 1: Direct double sandwich method.

This colour not only shows that antigen (virus) was present in the plant extract but also gives a quantitative measure of virus in each sample, since the amount of colour produced depends on the amount of enzyme in each well and this in turn depends on the degree of antigen–antibody binding.

The precision of this technique may be a disadvantage, since different strains of the same virus may not be detected. Such specificity may be useful in some situations, but for field tests where many strains of a given virus are present, it may mean many infected plants are not detected. To overcome this, an indirect ELISA method has been used with Andean potato latent virus (APLV) (Koenig, 1981). In this test, microtitre wells are first treated with antigen and then with antiserum produced in rabbits. The enzyme, however is conjugated to globulins that recognize rabbit serum; these are produced in chickens by injecting them with rabbit serum. This anti-rabbit globulin with its attached enzyme is added to the microtitre wells and then the substrate added (Fig. 7.8). This method lowered the specificity of the test and allowed the detection of a wider range of serologically related viruses. It was not suitable, however, for crude extracts since substances in the plant sap inhibit adsorption of the virus to the microtitre plates. Strains of APLV could be detected in crude saps

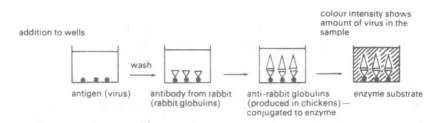

Figure 7.8 Enzyme linked immunosorbent assay (ELISA) tests. 2: Indirect method.

when indirect ELISA was performed on plates pre-coated with antiviral antibodies produced in rabbits. The trapped virus particles were then treated with a second antiviral antibody made by injecting virus into chickens. The bound chicken antibodies were finally detected using enzyme attached to globulins produced in rabbits that recognize chicken antibodies.

7.4.3 Electron microscope serology (or immunosorbent electron microscopy)

In these techniques virus and antiserum are reacted together and the results viewed in the electron microscope. Three main methods are employed:

(a) Grids are coated with antiserum, and virus-containing extract applied. Viruses reacting with the antiserum remain attached to the grid and can be detected by negative staining. This method can result in a 1000-fold or more increase in sensitivity over conventional electron microscopy in detecting viruses (Roberts and Harrison, 1979). It is probably the most sensitive method of all serological tests (Torrance and Jones, 1981). Results are rapid (1–4 hours) and use only small volumes of antisera. If required, grids can be treated with several antisera in order to detect several viruses at one time (Thomas, 1980).

(b) Viruses may be detected by 'decoration'. Virus-containing extract is applied to carbon-coated grids. The grid is treated with antiserum, and then negatively stained. Individual virus particles are examined for a 'halo' or decoration of antibody molecules. Viruses devoid of decoration are not antigenic to the serum used. This technique allows viruses in mixtures of similar-shaped and sized particles to be distinguished.

(c) In the third test, virus-containing sap is mixed with antiserum and a grid touched to the mixture. The grid is washed and then negatively stained. This method produces clumps of virus particles and also an approximately tenfold increase in numbers of particles on the grid (Milne and Luisoni, 1975). Virus particles not reacting with the antiserum do not clump together and are not coated with antibody molecules.

From these studies, information concerning virus shape, host range, effects on host, possible modes of transmission and serological relationships, if any, can be assessed. These data might be sufficient to justify the identification of the virus, but more precise information requires experimentation with purified virus.

7.5 Virus purification

Virus is preferably extracted from systemically infected hosts, since the virus concentration in these plants is generally higher than in local-lesion hosts.

Extraction of virus by grinding tissue using a pestle and mortar, or by sap press, is often better than vigorous mechanical homogenation, since elongated virus particles are broken by the latter method.

Cold buffer is the preferred extraction medium, the type of buffer and its pH varying from one extraction procedure to another, and the final choice depending largely on trial and error assessment. Most viruses are precipitated from solutions at acid pH, so it is best to use neutral pH when the viruses are most soluble. The ionic strength of the extracting medium is important; high ionic strength buffers may precipitate cell constituents and help in purification.

Buffers may contain reducing agents which prevent plant oxidase products from inactivating viruses. The most commonly used compounds are sodium sulphite, mercaptoethanol (one of the smelliest compounds on earth!), thioglycollic acid, and cysteine hydrochloride. Sodium diethyldi-thiocarbamate (DIECA) chelates copper and stops polyphenol oxidase enzymes. Tannins, present in plant extracts, may inactivate virus, but this can be prevented by adding 1 % nicotine.

Ethylene diamine tetra-acetic acid (EDTA) may be used to remove calcium and magnesium ions when they cause ribosomes to become unstable, thus removing them from the plant extract. To prevent nucleases breaking down viral nucleic acids during extraction, phenol or the diatomaceous earth bentonite may be added to the buffer.

Detergents (Tween 80, Triton X100) may be used during extraction to prevent aggregation of particles.

7.5.1 Precipitation of virus

If virus suspensions are treated with materials that either neutralize the natural changes on the virus particles or dehydrate those particles, then the virus will come out of solution. Thus, at a pH equivalent to the isoelectric point some viruses are reversibly precipitated. This property may be used for example to separate TNV (isoelectric point pH 4.5) from satellite virus (isoelectric point pH 7.0).

Ammonium sulphate can be used to both neutralize charges and to dehydrate virus although other proteins and ribosomes may be precipitated.

Dehydration by alcohol or acetone may be employed to precipitate cell materials leaving virus in solution. High alcohol concentration may destroy virus, although TMV for example can be reversibly precipitated unharmed by 50 % ethanol. Butanol, alone or mixed with chloroform, can be used to precipitate cellular material leaving virus in suspension. Butanol acts slowly and often the sap-butanol mixture is left for some hours before centrifugation to remove plant proteins. Ether, carbon tetrachloride or freon ($C \cdot Cl_2 F \cdot C \cdot ClF_2$) may be used instead of butanol.

Virus may be precipitated using the hydrophilic compound poly-ethylene glycol (PEG), MW 4000 or 6000 alone or with salt (NaCl). Precipitation depends on the virus and its concentration as well as the concentration of PEG and salt.

7.5.2 Centrifugation

Virus is often separated from cellular protein and debris by centrifugation. The simplest type of centrifugation is *differential centrifugation*. Since viruses are smaller than most cellular organelles, relatively low-speed centrifugation may be used to remove cell debris such as nuclei, mito-chondria, chloroplast and fragmented cell walls. Subsequent steps in purification may precipitate cell proteins and these, too, can be removed by centrifugation. Finally, the virus is removed from solution by high-speed centrifugation. Some viruses, such as TMV, can be purified by alternate cycles of low- and high-speed centrifugation. Care must be taken that virus particles are not damaged by pelleting.

Using *density gradient centrifugation* it is possible to obtain pure samples of virus and minimize damage. The virus sample is layered on to the top of a linear gradient of, say, 10–40% sucrose (Fig. 7.9). Centrifugation results in precipitation of the components of the virus preparation under the enhanced gravitational field at different rates to form layers in the tube dependent on their shape, size and density. This is termed *rate zonal centrifugation*.

If centrifugation is continued long enough, each component in the mixture will either sediment to the bottom of the tube, or if the density of the gradient is sufficient, each component of the mixture will float in the gradient at a point equal to its own density. This is called *equilibrium* or *isopycnic centrifugation*. Very often isopycnic centrifugation is carried out in a medium of caesium chloride or caesium sulphate. These compounds form natural gradients on centrifugation.

Extraction of virus, precipitation and centrifugation result in a purified

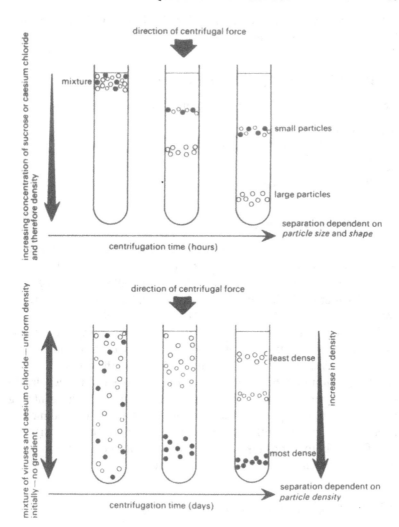

Figure 7.9 Density gradient centrifugation.

preparation of the virus, free, it is hoped, from plant cell or other contaminants. Pure virus preparations can be stored in a refrigerator at about 4°C, provided an antimicrobial substance such as chloroform or sodium azide is added. Alternatively, samples may be frozen, although freezing and thawing may disrupt virus particles. Some viruses can be lyophilized (freeze-dried).

Pure virus can now be used to establish physical and chemical characteristics of the virions.

7.6 Determination of physical characteristics

The characteristics most commonly quoted for plant viruses include:

(1) Sedimentation coefficient
(2) Partial specific volume and buoyant density
(3) Diffusion coefficient
(4) Particle molecular weight
(5) Absorbance ratio (A_{260}/A_{280}).

7.6.1   Sedimentation coefficient (S)

This is the velocity in centimetres per second at which a substance will sediment in a field of 10^{-5} newtons $(1/981\,g)$ in water at 20°C. A sedimentation coefficient of 1×10^{-13} seconds is called a *Svedberg* or $1S$. The sedimentation coefficient of TMV for example, is 194×10^{-13} sec. = $194S$, and that of the spherical virus TNV is about $118S$.

S is found by centrifugation of a virus preparation, using the equation $S = dx/dt/\omega^2 r$ where dx/dt = rate of sedimentation, ω = angular velocity in radians per second ($= 377$ rpm), and r = radius of the rotor.

Centrifugation may be carried out in a preparative centrifuge (Brakke, 1967; Kado and Agrawal, 1972) or in an analytical ultracentrifuge (Markham, 1967). In the latter, an optical system allows the sedimenting material to be viewed during centrifugation as a moving boundary (Fig. 7.10). From photographs taken of the boundary at known time intervals, the rate of sedimentation dx/dt can be calculated.

7.6.2   Partial specific volume and buoyant density

The partial specific volume $(\bar{v})$, the reciprocal of the buoyant density, is the volume of solvent, usually water or dilute salt solution, displaced by 1 g of virus when it is added to an infinite volume of solvent.

This volume is related to sedimentation coefficient (S) and molecular weight (M) by the equation

$$M = \frac{RTS}{D(1 - \bar{v}\rho)} \tag{1}$$

where R = gas constant, T = absolute temperature, D = diffusion coefficient, and ρ is the density of suspending medium, usually water.

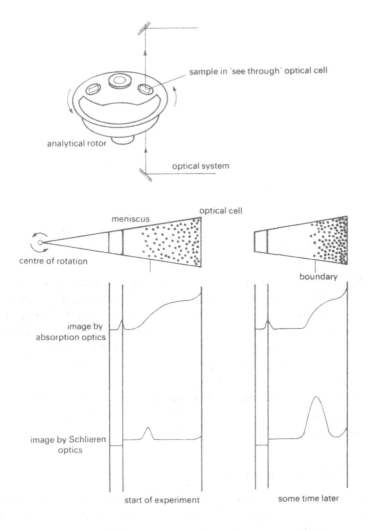

Figure 7.10 Analytical centrifugation—diagrammatic representation of its use in virus studies.

The partial specific volume is usually measured with a pycnometer (Fig. 7.11), a tiny vessel holding about 10 ml of fluid. The pycnometer is filled accurately with solvent and weighed at a constant known temperature, some of the solvent is removed, and a known weight of virus dissolved in the solvent is added. Pure solvent is added to refill the pycnometer accurately and the new weight ascertained.

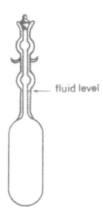

fluid level

Figure 7.11   Pycnometer—a small vessel of 10–12 ml volume for measuring the partial specific volume of virus suspensions (see text).

The difference between the increased weight and the weight of virus represents the amount of solvent displaced by the virus, and hence $\bar{v}$ can be determined. For example, let us assume that the solvent density at the temperature of the experiment is 1.02, and the difference in weight of the pycnometer containing 345 mg of virus was 115 mg more than that containing solvent alone. Then 345 mg of virus occupy $115/1.02 = 0.113\,cm^3$ *less* than 345 mg of pure solvent ($345/1.02 = 0.338\,cm^3$), i.e., $0.225\,cm^3$. This is the amount of solvent displaced by 0.345 g of virus, so that the partial specific volume $\bar{v} = 0.225/0.345 = 0.652\,cm^3\,g^{-1}$, and the buoyant density $= 1/0.652 = 1.533\,g\,cm^{-3}$.

7.6.3   Diffusion coefficient (D)

This is a measure of the amount of Brownian movement of a virus particle. It is a function of the kinetic energy of the particle, or its absolute temperature, and is inversely related to properties such as medium viscosity and particle size and shape. As we have seen in equation (1) above, diffusion coefficient is important in terms of measuring virus molecular weight. There are a number of ways of measuring D (Markham, 1967), but usually the method of free diffusion in one dimension is used. Briefly, an interface is made between pure solvent and solvent containing viruses. Diffusion at this interface is measured in a special apparatus by following refraction changes. The boundary between virus and solvent may be made in an analytical ultracentrifuge by initially spinning virus at high g forces into the centre of the centrifuge cell, and then decelerating to about 2500 rpm so that diffusion occurs at low g values.

7.6.4 Particle molecular weights

These can be found using equation (1) above—Table 4.3 lists some virus particle molecular weights.

7.6.5 Ultraviolet absorbance characteristics

Pure solutions of virus may appear opalescent since they scatter light, but they do not absorb visible light. Ultraviolet (UV) light, on the other hand, is absorbed in a characteristic way. The absorption of UV light by viruses is due both to their protein and their nucleic acid components. Absorbance by protein is mainly by the amino acids tryptophan and tyrosine, and shows maxima at 280 nm (Fig. 7.12) and a minimum at 260 nm. The nucleic acid bases, particularly adenine and uracil, absorb with maxima at 260 nm and minima at 230 nm (Fig. 7.13). Nucleic acid absorbance is about 20 times stronger than that by a similar concentration of protein. These features of UV light absorption are very convenient in virus analysis.

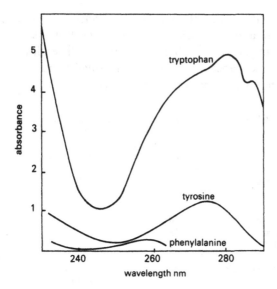

Figure 7.12 Absorbance of amino acids contributing to the major absorbance of protein in UV light.

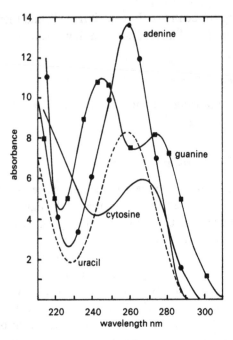

Figure 7.13 Absorbance of nucleic acid bases in UV light.

Each virus has a fixed ratio of protein to nucleic acid; TMV for example is 95% protein and 5% nucleic acid by weight. If it is scanned with UV light wavelengths then absorbance at 280 nm will result from its protein content and absorbance at 260 nm from its smaller (but more absorptive) RNA content. Thus the ratio of absorbance at 260 nm (A_{260}) to

Table 7.2 Ultraviolet absorbance characteristics of plant viruses.

Virus	Percentage nucleic acid	A_{260}/A_{280}	$E_{1\,cm}^{0.1}$
Tobacco mosaic	5	1.19	2.7–3.5
Potato virus X	6	1.20	2.97
Potato virus Y	6	1.21	2.30
Cucumber mosaic	18	1.7	5.0
Tobacco necrosis	19	1.7	5.0–5.5
TNV satellite	20	1.7	6.5
Bean golden mosaic*	29	1.4	7.7
Turnip yellow mosaic	35	0.81 (T)	0.96 (T)
		1.51 (B)	9.60 (B)

* DNA geminate virus.

absorbance at 280 nm (A_{280}) has a value of 1.19 for TMV and is a fixed feature of that virus. Table 7.2 shows the A_{260}/A_{280} ratio and percentage RNA for several viruses. Viruses with high protein content will have lower A_{260}/A_{280} ratio than those with more nucleic acid.

If working with a known purified virus, a ratio differing from that expected may indicate the presence of impurities; values lower than expected may indicate contamination by host proteins, for instance. In addition, some apparent increase in absorbance comes about because light is scattered by the viruses in suspension; for this reason absorbance needs to be corrected. Light scattering in the UV cannot be measured directly but is extrapolated from figures obtained in visible light (Fig. 7.14).

As well as helping in identifying virus and in assessing purity, absorbance measurements can be used to estimate concentrations. Initially a purified virus preparation can be dried and weighed on a very precise micro-balance with an accuracy of 0.002 mg. A solution of virus of known strength can be prepared and absorbance measured at 260 nm. Usually

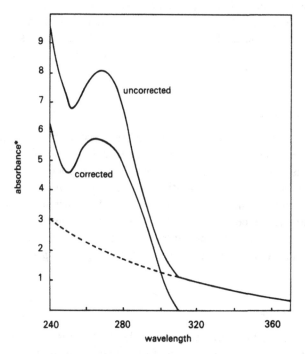

Figure 7.14 Absorbance spectrum of tobacco mosaic virus. *Absorbance extrapolated from 600–315 nm readings. (1) Uncorrected. (2) Corrected (* values deducted from 1).

absorbance is measured for a 1 cm layer of a 1 mg/ml virus preparation, this being termed the *extinction coefficient* (E), written as $E_{1\,cm}^{0.1\%}$. Some published E values are given in Table 7.2.

7.7   Determination of chemical characteristics

Chemical analysis can be used to show the type of nucleic acid present, the nature of the protein component, and the presence, if any, of other chemical substances.

7.7.1   *Analysis of nucleic acid*

This involves establishing (a) whether the virus contains RNA or DNA, (b) whether the nucleic acid is single- or double-stranded, (c) details of base sequence and (d) some idea of the nucleic acid molecular weight.

    Extraction of nucleic acid is achieved by a variety of methods, but the most usual is that using saturated aqueous phenol, either alone or together with bentonite, or the detergent sodium dodecyl sulphate (SDS). In other procedures SDS may be used alone or with sodium perchlorate. The type of nucleic acid is determined by its base composition, by its sensitivity to either DNase or to RNase, or by its buoyant density (Hamilton *et al.*, 1981).

7.7.2   *Strandedness*

In double-stranded nucleic acid, base pairing occurs between guanosine and cytidine and between adenosine and either thymidine (DNA) or uracil (RNA). It is possible to separate the two strands by chemical treatment or by heating. When heated, separation or 'melting' occurs at a temperature which is found by following changes in the UV absorbance. The melting temperature varies between 85–100°C and depends on the ionic strength and the guanine and cytosine content (% GC) of the nucleic acid. Single-stranded nucleic acids do not show this melting, or *hyperchromic*, effect (Fig. 7.15).

7.7.3   *Base sequence*

This is determined essentially by controlled hydrolysis of the nucleic acid using enzymes, and the separation of the products by electrophoresis and chromatography. If the products of hydrolysis at various stages are

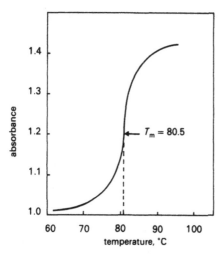

Figure 7.15 Hyperchromic effect of raising the temperature of double-stranded nucleic acid. The melting temperature T_m is the mid-point of increase in absorbance.

examined, it is possible to determine the composition of fragments and to work out how these fit together and so establish the sequence of the complete molecule. More recently, nucleic acid hybridization techniques have been used. Comparison is made between the degree of homology between reference negative-strand RNA or reference complementary DNA molecules.

7.7.4 MW determinations

Separated nucleic acid can be centrifuged and molecular weights determined from the movement of preparations in the ultracentrifuge. RNA molecules can be separated on polyacrylamide or agarose gels (Fig. 7.16). RNA is located with toluidine blue dye or under UV light after staining with ethidium bromide. However, extensive secondary structure of single-stranded RNA molecules may make molecular weight determinations inaccurate. To overcome this problem, RNA may be denatured by treatment with methyl mercuric hydroxide, a highly toxic material, or more safely with 8M urea or glyoxal, and the resulting RNA electrophoresed in 0.75% agarose gels. Mobilities show a straight line relationship to log molecular weight (Murant *et al.*, 1978, 1981). By comparing with standard RNAs of known molecular weight, unknowns may be characterized.

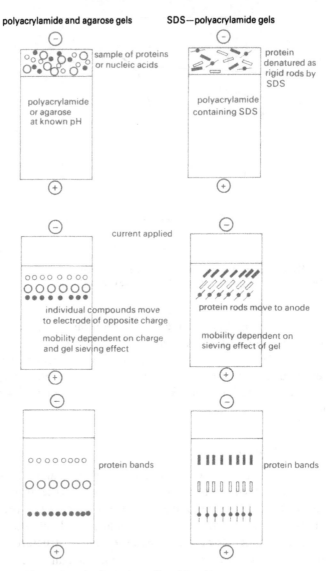

Figure 7.16 Electrophoresis of proteins and nucleic acids.

7.7.5 Protein

The protein of plant viruses can be isolated by mixing purified virus preparations with water-saturated phenols. The mixture is centrifuged to break up the emulsion, the nucleic acid being in the aqueous phase and the protein either precipitated or present in the phenol. The phenol is removed by washing with methanol. Such protein is not 'native' protein in the sense that it will reconstitute in viral protein shells; it is, however, suitable for further chemical analysis. Other techniques used to isolate protein from virions include alkaline treatment, or vigorous treatment with trichloro-acetic acid, acetic acid or formic acid. Some plant virus proteins can be obtained by brief exposure to high temperature (100°C) in the presence of salt for five minutes or more. Milder heat treatment (45°C for 30 minutes) at pH 8.0 in the presence of 0.1 M NaCl, in turnip yellow mosaic virus for example, yields protein that can be separated from RNA by centrifugation. Sodium dodecyl sulphate can also be used to dissociate a number of viruses.

The molecular weight of proteins is most commonly found by gel electrophoresis in the presence of the denaturing detergent SDS. Proteins denatured by SDS are fairly uniformly coated with detergent, and show a uniform dependence of electrophoretic mobility in polyamylamide gels on molecular weight. The molecular weight of the viral protein is determined by comparison with the migration of a set of marker proteins.

If time, equipment, expertise and money are available, then starting from our diseased material we could arrive at a purified virus sample with enough information to write an adequate description of the virus similar to those published by the CMI/AAB. We could also produce a cryptogram (section 1.6), placing the virus either into a recognized group or possibly in new virus groups.

BIBLIOGRAPHY

General section

The following are general texts, reviews and articles on special topics used in the preparation of the text and relevant to the whole of plant virology. Useful reviews are found regularly in:

Advances in Virus Research
Annual Reviews of Biochemistry
Annual Reviews of Microbiology
Annual Reviews of Phytopathology
Intervirology

Anon. *Crop Protection in Northern Britain*. Proceedings of a conference held at Dundee University, March 1981.

Bawden, F. C. (1956) *Plant Viruses and Virus Diseases*. 3rd edn., Chronica Botanica Co., Waltham, Mass.

Beemster, A. B. R. and Dijkstra, J. (1966) *Viruses of Plants*. North Holland Pub. Co., Amsterdam.

Bos, L. (1970) *Symptoms of Virus Diseases of Plants*. 2nd edn., PUDOC, Wageningen.

Carter, W. (1973) *Insects in Relation to Plant Disease*. 2nd edn., John Wiley and Sons, New York.

CMI/AAB (1970–1981) *Descriptions of Plant Virus*. Commonwealth Mycological Institute and Association of Applied Biologists, Kew.

Cooper, J. I. (1979) *Virus Diseases of Trees and Shrubs*. Institute of Terrestrial Ecology.

Corbett, M. K. and Sisler, H. D. (1964) *Plant Virology*. University of Florida Press, Gainesville.

Ebbels, D. L. and King, J. E. (1980) *Plant Health. The Scientific Basis for Administrative Control of Plant Diseases and Pests*. Blackwell Scientific Publications, Oxford.

Edwardson, J. R. and Christie, I. R. G. (1978) Use of virus-induced inclusions in classification and diagnosis. *Am. Rev. Phytopath.* **16**, 31–55.

Esau, K. (1968) *Viruses in Plant Hosts: Distribution and Pathogenic Effect*. University of Wisconsin Press.

Fraenkel-Conrat, H. (1969) *The Chemistry and Biology of Viruses*. Academic Press, New York.

Fraenkel-Conrat, H. and Wagner, R. R. *Comprehensive Virology*. Vols. 1 onwards. Plenum Press, New York.

Francki, R. I. B. (1973) Plant rhabdoviruses. *Adv. Vir. Res.* **18**, 257–345.

Fridlund, P. R. (1980) The 1R-2 program for obtaining virus-free fruit trees. *Plant Disease*, **64**, 826–30.

Gibbs, A. J. (ed.) (1973) *Viruses and Invertebrates.* North Holland Pub. Co., Amsterdam.

Gibbs, A. and Harrison, B. D. (1976) *Plant Virology. The Principles.* Edward Arnold.

Harrison, B. D. (1980) A biologist's view of 25 years of plant virus research. *Ann. appl. Biol.* **94**, 321–33.

Harrison, B. D. (1981) Plant virus ecology: ingredients, interactions and environmental influences. *Ann. appl. Biol.* **99**, 195–209.

Harrison, B. D., Finch, J. T., Gibbs, A. J., Holling, M., Shepherd, R. J., Valenta, V. and Wetter, C. (1971) Sixteen groups of plant viruses. *Virology* **45**, 356–63.

Horne, R. W. (1974) *Virus Structure.* Academic Press, New York.

Ingram, D. S. and Helgeson, J. B. (1980) *Tissue Culture Methods in Plant Pathology.* Blackwell Scientific Publications, Oxford.

Kado, C. I. and Agrawal, H. O. (1972) *Principles and Techniques in Plant Virology.* Van Nostrand, Reinhold & Co.

Kassanis, B. (1980) Therapy of virus infected plants and the active defence mechanism. *Outlook on Agric.* **10**, 288–92.

Luria. D. E., Darnell, J. E., Baltimore, D. and Campbell, A. (1978) *General Virology.* 3rd edn., John Wiley and Sons, New York.

Maramorosch, K. (ed.) (1962) *Biological transmission of disease agents.* Academic Press, New York.

Maramorosch, K. (1969) *Viruses, Vectors and Vegetation.* John Wiley and Sons, New York.

Maramorosch, K. and Harris, K. F. (1981) *Plant Diseases and Vectors, Ecology and Epidemiology.* Academic Press, New York.

Maramorosch, K. and Koprowski, P. (1967) *Methods in Virology,* Vol. II. Academic Press, New York.

Martin, H. (1971) *Pesticide Manual.* 2nd edn., British Crop Protection Council.

Martyn, E. B. (ed.) (1968) *Plant Virus Names.* Phytopathological Paper No. 9. Kew, CMI.

Martyn, E. B. (ed.) (1971) *Ibid.,* Supplement No. 1.

Matthews, R. E. F. (1957) *Plant Virus Serology.* Cambridge University Press.

Matthews, R. E. F. (1970) *Plant Virology.* Academic Press, London and New York.

Matthews, R. E. F. (1981) *Plant Virology.* 2nd edn., Academic Press, London and New York.

Matthews, R. E. F. (1979) Classification and nomenclature of viruses. *Intervirology* **12**, 132–296.

Noordam, D. (1973) *Identification of Plant Viruses. Methods and Experiments.* PUDOC, Wageningen.

Romberg, J. A. (ed.) (1977) *Virology in Agriculture.* Beltsville Symposia in Agricultural Research, No. 1.

Russell, G. (1978) *Plant Breeding for Pest and Disease Resistance.* Butterworth, London.

Schneider, H. (1973) Cytological and histological aberrations in woody plants following infection with viruses, mycoplasmas, rickettsias and flagellates. *Ann. Rev. Phytopath.* **11**, 119–46.

Scott, P. R. and Bainbridge, A. (1978) *Plant Disease Epidemiology.* Edit. for Fed. of British Plant Path., Blackwell Scientific Publications, Oxford.

Simons, J. N. and Zittel, T. A. (1980) Use of oil to control aphid-borne virus. *Plant Disease* **64**, (6), 542–6.

Slack, S. A. (1980) Pathogen-free plants by meristem-tip culture. *Plant Disease* **64**, 14–17.

Smith, K. M. (1972) *A Text-book of Plant Virus Diseases.* 3rd edn., Longman, London.

Thresh, J. M. (1974) Vector relationships and the development of epidemics: the epidemiology of plant viruses. *Phytopath.* **64**, 1050–6.

Voller, A. Bidwell, D. E. and Barlett, A. (1977) *The Enzyme Linked Immunosorbent Assay (ELISA).* Flowline Publication.

Walters, H. J. (1969) Beetle transmission of plant viruses. *Adv. Virus Res.* **15**, 339–63.

Waterson, A. P. and Wilkinson, L. (1978) *An Introduction to the History of Virology.* Cambridge University Press.

Chapter references

Chapter 1
Anon (1979) ADAS Annual Report. H.M.S.O., London.
Bawden, F. C. (1956) *Plant Viruses and Virus Diseases.* 3rd edn., Chronica Botanica Co., Waltham, Mass.
Gibbs, A., Harrison, B. D., Watson, D. H. and Wildy, P. (1966) What's in a virus name? *Nature* **209**, 450–4.
Gibbs, A. and Harrison, B. D. (1968) Realistic approach to virus classification and nomenclature. *Nature* **218**, 927–9.
Gibbs, A. and Harrison, B. D. (1976) *Plant Virology. The Principles.* Edward Arnold.
Heathcote, G. D. (1978) Effects of virus yellows on yield of some monogerm cultivars of sugar beet. *Ann. appl. Biol.* **88**, 145–51.
Inouye, T. and Osaki, T. (1980) The first record in the literature of the possible plant virus disease that appeared in 'Manyoshu', a Japanese classic anthology, as far back as the time of the 8th Century. *Ann. Phytopath. Soc. Jap.* **46**, 49–50.
Legg, J. T. (1979) The campaign to control the spread of cocoa swollen shoot virus in Ghana, in Ebbels and King (1979)—see general section.
Luria, D. E., Darnell, J. E., Baltimore, D. and Campbell, A. (1978) *General Virology.* 3rd edn., John Wiley and Sons, New York.
Martyn, E. B. (ed.) (1968) Plant Virus Names. Phytopathological Paper No. 9. Kew, CMI.
Martyn, E. B. (ed.) (1971) Plant Virus Names. *Ibid.*, supplement No. 1.
Matthews, R. E. F. (1970) Classification and Nomenclature of Viruses. Intervirology **12**, 132–296.
Thresh, J. M. and Pitcher, R. S. (1978) In Scott and Bainbridge (1978)—see general section.
Uyemoto, J. K., Claflin, L. E., Wildon, D. L. and Raney, R. J. (1981) Maize chlorotic mottle and maize dwarf mosaic, effect of single and double inoculation on symptomatology and yield. *Plant Disease* **65**, 39–40.
Villalon, B. (1981) Breeding peppers to resist virus diseases. *Plant Disease* **65**, 557–562.

Chapter 2
Akimoto, T. A., Wagner, M. A., Johnson, J. E. and Rossmann, M. G. (1975) The packing of southern bean mosaic virus in various crystal cells. *J. Ultrastruct. Res.* **53**, 306–18.
Andrews, J. H. and Shalla, T. A. (1974) The origin, development and conformation of amorphous inclusion body components in tobacco etch virus-infected cells. *Phytopath.* **64**, 1234–43.
Atkinson, P. H. and Matthews, R. E. F. (1967) Distribution of tobacco mosaic virus in systemically infected tobacco leaves. *Virology* **32**, 171–3.
Baker, E. A. and Campbell, A. I. (1966) Effects of viruses on pigment composition of apple-bark and pear leaf. *Ann. Rep. Long Ashton Res. Stn.*, 141–4.
Christie, R. G. and Edwardson, J. R. (1977) *Light and Electron Microscopy of Plant Virus Inclusions.* Florida Agric. Exptl. Stn. Monograph No. 9.
Cronshaw, J., Hoefert, L. and Esau, K. (1966) Ultrastructural features of *Beta* leaves infected with beet yellows virus. *J. Cell Biol.* **31**, 429–43.
Edwardson, J. R. (1974) *Some Properties of the Potato Virus Y Group.* Florida Agric. Exptl. Stn. Monograph No. 4.
Edwardson, J. R. (1974) *Host Ranges of Viruses in the BVY Group.* Florida Agric. Exptl. Stn. Monograph No. 5.
Esau, K. (1956) An anatomist's view of virus diseases. *Am. J. Bot.* **43**, 739–48.
Esau, K. (1968) *Viruses in Plant Hosts: Distribution and Pathogenic Effect.* University of Wisconsin.
Esau, K. and Cronshaw, J. (1967) Relation of TMV with host cells. *J. Cell Biol.* **33**, 665–78.

Francki, R.I.B. and Randles, J.W. (1970) *Lettuce Necrotic Yellow Virus.* CMI/AAB Descriptions of Plant Virus, No. 26.

Goldstein, B. (1924)—see Martelli and Russo, (1977).

Hatta, T. (1976) Recognition and measurement of small isometric virus particles in thin sections. *Virology* **69**, 237–45.

Hatta, T., Bullivant, S. and Matthews, R. E. F. (1973) Fine structure of vesicles in chloroplasts of Chinese cabbage leaves by infection with turnip yellow mosaic virus. *J. Gen. Virol.* **20**, 37–50.

Hatta, T. and Matthews, R. E. F. (1974) The sequence of early cytological changes in Chinese cabbage leaf cells following systemic infection with turnip yellow mosaic virus. *Virology* **59**, 383–6.

Herbert, T. T. and Panizo, C. H. (1975) *Oat Mosaic Virus.* CMI/AAB Descriptions of Plant Virus No. 145.

Holmes, F. O. (1929) Local lesions in tobacco mosaic. *Bot. Gaz.* **87**, 39–55.

Holmes, F. O. (1964)—see Corbett and Sisler (1964), general section.

Hull, R., Hills, G.J. and Plaskitt, A. (1970) The *in vivo* behaviour of twenty-four strains of alfalfa mosaic virus. *Virology* **42**, 753–72.

Iwanowski, D. (1903) Über die mosaikkrankheit der tabakspflanze. *Z. Pfl. Krankh.* **13**, 1–41.

Johnson, J. E., Rossmann, M. G., Smiley, I. E. and Wagner, M. A. (1974) Single crystal X-ray diffraction studies of southern bean mosaic virus. *J. Ultrastruct. Res.* **46**, 441–51.

Knight, R. and Tinsley, T. W. (1958) Some histological observations on virus-infected *Theobroma cacao* L. *Ann. appl. Biol.* **46**, 7–10.

Langenberg, W. G. (1979) Chilling of tissue before glutaraldehyde fixation preserves fragile inclusions of several plant viruses. *J. Ultrastruct. Res.* **66**, 120–31.

Langenberg, W. G. and Schroeder, H. F. (1975) The ultrastructural appearance of cowpea mosaic virus in cowpea. *J. Ultrastruct. Res.* **51**, 166–75.

Lawson, R. H., Hearon, S. S. and Civerolo, E. L. (1977) *Carnation Etched Ring Virus.* CMI/AAB Descriptions of Plant Viruses No. 182.

Martelli, G. P. and Russo, M. (1977) Plant virus inclusion bodies. *Adv. in Virus Res.* **21**, 175–256.

Martin, C. (1958) Anomalies de synthèse des anthocyanes dans le genre de pomme de terre atteinte de maladies à virus. *Comp. rend. hebd. Séanc Acad. Sci. Paris* **246**, 2790–2.

Matsui, C. (1959) Fine Structure of X-body. *Virology* **9**, 306–13.

Mayer, A. (1886) Über die Mosaikkrankheit des Tabaks. *Landwn. Vers. Stnen.* **32**, 450–67.

Reid, M. W. and Matthews, R. E. F. (1966) On the origin of the mosaic induced by turnip yellow mosaic virus. *Virology* **28**, 563–70.

Shalla, T. A. (1964) Assembly and aggregation of TMV in tomato leaflets. *J. Cell Biol.* **21**, 253–64.

Shepardson, S. and McCrum, R. (1980) Extracytoplasmic tubules in leafroll infected and leafroll-free potato leaf tissue. *J. Ultrastruct. Res.* **72**, 47–51.

Tepfer, S. S. and Chessin, M. (1959) Effects of tobacco mosaic virus on early leaf development in tobacco. *Am. J. Bot.* **46**, 496–509.

Warmke, H. E. (1969) A reinterpretation of amorphous inclusions in the aucuba strain of TMV. *Virology* **39**, 695–704.

Chapter 3

Bradley, R. H. E. (1952) Studies on the aphid transmission of a strain of henbane mosaic virus. *Ann. appl. Biol.* **39**, 78–97.

Bradley, R. H. E. (1961) *Recent Advances in Botany*, Univ. of Toronto Press, Vol. 1, pp. 528–33.

Broadbent, L. and Fletcher, J. T. (1963) The epidemiology of tomato mosaic. IV. Persistence of virus on clothing and glasshouse structures. *Ann. appl. Biol.* **52**, 233–41.

Cadman, C. H. (1963) Biology of soil-borne viruses. *Ann. Rev. Phytopath.* **1**, 143–72.

Cockbain, A. J., Cook, S. A. and Bowen, R. (1975) Transmission of broad bean stain virus and echtes ackerbohnenmosaik-virus to field beans (*Vicia faba*) by weevils. *Ann. appl. Biol.* **81**, 331–9.

Dixon, A. F. G. (1973) *Biology of Aphids.* Edward Arnold.

Garrett, R. G. (1973) Non-persistent aphid borne viruses, in A. J. Gibbs (ed.) (1973) *Viruses and Invertebrates*, North-Holland Publ. Co., Amsterdam.

Garrett, R. G. (1971)—see Garrett (1973).

Harris, K. F. and Bath, J. E. (1973) Regurgitation of *Myzus persicae* during membrane feeding—its likely function in non-persistent plant viruses. *Animal Entomol. Soc. of Amer.* **66**, 793–6.

Hewitt, W. B., Raski, D. J. and Goheen, A. C. (1958) Nematode vector of soil-borne fanleaf virus of grape vines. *Phytopath.* **48**, 586–95.

Hunter, J. A., Chamberlain, E. E. and Atkinson, J. D. (1958) Note on transmission of apple mosaic by natural root grafts. *N.Z. Agr. Res.* **1**, 80–2.

Kassanis, B. and Govier, D. A. (1971) The role of the helper virus in aphid transmission of potato aucuba mosaic virus and potato virus C. *J. Gen. Virol.* **13**, 221–8.

McKay, M. P. and Warner, M. F. (1933) Historical sketch of tulip mosaic or breaking. The oldest known plant virus disease. *Natn. Hort. Mag.*, p. 187.

Murant, A. F. (1978) Recent studies on association of two plant virus complex with aphid vectors, in Scott and Bainbridge (1978)—see general section.

Nambra, R. (1962) Aphid transmission of plant viruses from the epidermis and sub-epidermal tissues; *Myzus persicae* (Sulzer) cucumber mosaic virus. *Virology* **16**, 267–70.

O'Loughlin, G. T. and Chambers, S. C. (1967) The systemic infection of an aphid by a plant virus. *Virology* **33**, 262–71.

Peters, D. and Black, L. M. (1970) Infection of primary cultures of aphid cells with a plant virus. *Virology* **40**, 847–53.

Richardson, J. and Sylvester, E. S. (1965) Aphid honeydew as inoculum for the injection of pea aphids with pea-enation virus. *Virology* **25**, 472–5.

Roberts, F. M. (1940) Studies on the feeding methods and penetration rates of *Myzus persicae* (Sulz.), *Myzus circumflexus* Buck. and *Macrosiphum gei* Koch. *Ann. appl. Biol.* **27**, 348–58.

Smith, K. M. (1946) Tobacco rosette: a complex virus disease. *Parasitology* **37**, 131–4.

Sylvester, E. S. (1954) Aphid transmission of non-persistent plant viruses with special reference to the *Brassica nigra* virus. *Hilgardia* **23**, 53–98.

Sylvester, E. S. (1962) Virus transmission by aphids—a viewpoint, in Maramorosch (ed.) (1962)—see general section.

Sylvester, E. S. (1969) Evidence of transovarial passage of the sowthistle yellow vein virus in the aphid *Hyperomyzus lactucae*. *Virology* **38**, 440–6.

Sylvester, E. S. and Richardson, J. (1969) Additional evidence of multiplication of the sowthistle yellow vein virus in an aphid vector—serial passage. *Virology* **37**, 26–31.

Sylvester, E. S. and Richardson, J. (1970) Infection of *Hyperomyzus lactucae* by sowthistle yellow vein virus. *Virology* **42**, 1023–42.

Thomas, W. D. Jr. and Baker, R. R. (1952), in Corbett and Sisler (1964)—see general section.

Watson, M. A. and Roberts, F. M. (1939) A comparative study of the transmission of *Hyoscyamus* virus 3, potato virus Y and cucumber virus 1 by the vectors *Myzus persicae* (Sulz.), *M. circumflexus* (Buckton) and *Macrosiphum gei* (Koch). *Proc. Roy Soc. London B*, **127**, 543–76.

Watson, M. A. and Roberts, F. M. (1940) Evidence against the hypothesis that certain plant viruses are transmitted mechanically by aphids. *Ann. appl. Biol.* **27**, 227–33.

Chapter 4

Best, R. (1968) Tomato spotted wilt virus. *Adv. Virus Res.* **13**, 66–146.

Bouley, J. P., Briand, J. P. Genevaux, M., Pinck, M. and Witz, J. (1976) The structure of eggplant mosaic virus. Evidence for the presence of low MW RNA in top component. *Virol.* **69**, 775–81.

Casper, D. L. D. (1956) Structure of bushy stunt virus. *Nature* **177**, 475–6.

Casper, D. L. D. (1964) Structure and function of regular virus particles, in Corbett and Sisler (1964)—see general section.

Casper, D. L. D. and Klug, A. (1962) Physical principles in the construction of regular viruses. *Cold Spring Harbor Symp. Quant. Biol.* **27**, 1–24.

Cohen, S. S. and McCormick, F. P. (1979) Polyamines and virus multiplication. *Adv. Virus Res.* **24**, 331–87.

Crick, F. H. C. and Watson, J. D. (1956) Structure of small viruses. *Nature* **177**, 473–5.

Egbert, L. N., Egbert, L. D. and Mumford, D. L. (1976) Physical characteristics of sugar beet curly top virus. *Abs. Ann. Meeting Am. Soc. Microbiol*, p. 258.

Francki, R. I. B., Hatta, T., Boccardo, G. and Randles, J. W. (1980) The composition of *Chloris* striate mosaic virus, a gemini virus. *Virology* **101**, 233–41.

Fulton, R. W. (1968), in Gibbs and Harrison (1976)—see general section.

Goodman, R. M., Shock, T. L., Haber, S., Browning, K. S. and Bowers, G. R. Jr. (1980) The composition of bean golden mosaic virus and its single stranded DNA genome. *Virology* **106**, 168–72.

Hahn, P. and Shepherd, R. J. (1980) Phosphorylated proteins in cauliflower mosaic virus. *Virology* **107**, 295–7.

Hatta, T. and Francki, R. I. B. (1977) Morphology of Fiji disease virus. *Virology* **76**, 797–807.

Heijtink, R. A., Houwing, C. J. and Jaspars, E. M. J. (1977) MW of particles and RNAs of alfalfa mosaic virus. Number of units in protein capsids. *Biochem.* **16**, 4684–93.

Jackson, A. O. and Christie, S. R. (1979) *Sonchus Yellow Net Virus*. CMI/AAB Description of Plant Viruses, No. 205.

Jeske, H. and Werz, G. (1980) Cytochemical characterization of plastidal inclusions in abutilon mosaic infected *Malva parviflora* mesophyll cells. *Virology* **106**, 155–8.

Klug, A., Finch, J. T. and Franklin, R. M. (1957) The structure of turnip yellow mosaic virus. 1. X-ray diffraction studies. *Biochim. Biophys. Acta* **25**, 242–52.

Milne, R. G. (1967) Plant viruses inside cells. *Sci. Prog.* **55**, 203–22.

Partridge, J. E., Shannon, L. M., Gumpf, D. J. and Colbaugh, P. (1974) Glycoprotein in the capsid of plant viruses as a possible determinant of seed transmissibility. *Nature* **247**, 391–2.

Richardson, J. F., Tollin, P. and Bancroft, J. B. (1981) Architecture of the potexviruses. *Virology* **112**, 34–39.

Van Vloten-Doting, L., Bol, J. F., and Jaspars, E. M. J. (1977), in Romberg, J. A. (ed.) (1977)—see general section.

Varma, A., Gibbs, A. J., Wood, R. D. and Finch, J. T. (1968) Some observations on the structure of the filamentous particles of several plant viruses. *J. Gen. Virol.* **2**, 107–14.

Wilson, T. M. A., Perham, R. N., Finch, J. T. and Butler, P. J. G. (1976) Polarity of the RNA in tobacco mosaic virus particles and the direction of protein stripping in sodium dodecyl sulphate. *FEBS Lett.* **64**, 285–9.

Chapter 5

Atabekov, J. G. and Morozov, S. Yu. (1979) Translation of plant virus messenger RNAs. *Adv. Virus Res.* **25**, 1–91.

Baltimore, D. (1971) Expression of animal virus genomes. *Bacteriol. Rev.* **35**, 235–41.

Bancroft, J. B., McLean, G. D., Rees, M. W. and Short, M. N. (1971) The effect of an arginyl to a cysteinyl replacement in the uncoating behaviour of a spherical plant virus. *Virology* **45**, 707–15.

Bawden, F. C. and Harrison, B. D. (1955) Studies on the multiplication of a tobacco necrosis virus in inoculated leaves of French-bean plants. *J. Gen. Microbiol.* **13**, 494–508.

Black, D. R. and Knight, C. A. (1970) Ribonucleic acid transcriptase activity in purified wound tumour virus. *J. Virol.* **6**, 194–8.

Bisaro, D. M. and Siegel, A. (1980) A new viral RNA species in tobacco rattle virus-infected tissue. *Virology* **107**, 194–201.

Brants, D. H. (1964) The susceptibility of tobacco and bean leaves to TMV infection in relation to the condition of the ectodesmata. *Virology* **23**, 588–94.

Butler, P. J. G. and Durham, A. C. H. (1977) Tobacco mosaic virus protein aggregation and the virus assembly. *Adv. Prot. Chem.* **31**, 187–251.

Cocking, E. C. (1966) An electron-microscopic study of the initial stages of infection of isolated tomato fruit protoplasts by tobacco mosaic virus. *Planta* **68**, 206–14.

Cocking, E. C. and Pojnar, E. (1969) An electron microscope study of the infection of isolated tomato fruit protoplast by tobacco mosaic virus. *J. Gen. Virol.* **4**, 305–12.

Covey, S. N. and Hull, R. (1981) Transcription of cauliflower mosaic virus DNA. Detection of transcript properties and location of the gene encoding the virus inclusion body protein. *Virology* **111**, 463–74.

Daubert, S. D., Bruening, G. and Najarian, R. C. (1978) Protein bound to the genome RNAs of cowpea mosaic virus. *Eur. J. Biochem.* **92**, 45–51.

El Manna, M. M. and Bruening, G. (1973) Polyadenylate sequences in the ribonucleic acids of cowpea mosaic virus. *Virology* **56**, 198—206.

Erikson, J. W. and Bancroft, J. B. (1981) Melting of viral RNA by coat protein: assembly strategies for elongated plant viruses. *Virology* **108**, 235–40.

Francki, R. I. B. (1973) Plant rhabdoviruses. *Adv. in Virus Res.* **18**, 257–345.

Fukada, M., Ohno, T., Okada, Y., Otuski, Y. and Takebe, I. (1978) Kinetics of biphasic reconstitution of tobacco mosaic *in vitro*. *Proc. Nat. Acad. Sci. USA* **75**, 1727–30.

Gaard, G. and DeZoeten, G. A. (1979) Plant virus uncoating as a result of virus cell wall interactions. *Virology* **96**, 21–31.

Gerola, F. M., Bassi, M., Favali, M. A. and Betto, E. (1969) An electron microscope study of the penetration of tobacco mosaic virus into leaves following experimental inoculation. *Virology* **38**, 380–6.

Goodman, R. M. (1981) Gemini viruses. *J. Gen. Virol.* **54**, 9–21.

Haber, S., Ikegami, M., Bajet, N. B. and Goodman, R. M. (1981) Evidence for a divided genome in bean golden mosaic virus, a gemini virus. *Nature* **289**, 324–6.

Harrison, B. D. and Robinson, D. J. (1978) The tobraviruses. *Adv. Virus Res.* **23**, 25–77.

Hull, R. and Shepherd, R. J. (1977) The structure of cauliflower mosaic virus genome. *Virology* **79**, 216–30.

Hunter, T. R., Hunt, T., Knowland, J. and Zimmern, D. (1976) Messenger RNA for the coat protein of TMV. *Nature* **260**, 759–64.

Jackson, A., Mitchell, D. and Siegel, A. (1971) Replication of TMV. I. Isolation and characterization of double-stranded forms of ribonucleic acid. *Virology* **45**, 182–91.

Kassanis, B. and Kenton, R. H. (1978) Inactivation and uncoating of TMV on the surface and intercellular spaces of leaves. *Phytopath Z.* **91**, 329–39.

Kurtz-Fritsch, C. and Hirth, L. (1972) Uncoating of two spherical plant viruses. *Virology* **47**, 385–96.

Martin, S. J. (1978) *The Biochemistry of Viruses*. Cambridge University Press.

Merkens, W. S. W., De Zoeten, G. A. and Gaard, G. (1972) Observation on ectodesmata and the virus infection process. *J. Ultrastruct. Res.* **41**, 397–405.

Niblett, C. L. (1972) Attachment of plant viruses to host and non-host plants. *Proc. Amer. Phytopath. Soc.* **2**, 87.

Okuno, T. and Furusawa, I. (1979) RNA polymerase activity and protein synthesis in brome mosaic virus infected protoplasts. *Virology* **99**, 218–25.

Pelham, H. R. B. (1979) Translation of tobacco rattle virus RNAs *in vitro*. Four proteins from 3 RNAs. *Virology* **97**, 256–65.

Razelman, G., Goldbach, R. and Van Kammen, A. (1980) Expression of bottom component

RNA of cowpea mosaic virus in cowpea protoplasts. *J. Virol.* **36**, 366–73.

Rhodes, D. P., Reddy, D. V. R., Macleod, R., Black, L. M. and Bannerjee, A. K. (1977) *In vitro* synthesis of RNA containing 5'-terminal structure ^{7}m G(5') ppp (5') A_p^m ... by purified wound tumour virus. *Virology* **76**, 554–9.

Shaw, J. G. (1969) *In vivo* removal of protein from tobacco mosaic virus after inoculation of tobacco leaves. II. Some characteristics of the reaction. *Virology* **37**, 109–16.

Shepherd, R. J. (1976) DNA viruses of higher plants. *Adv. Virus Res.* **20**, 305–39.

Siegel, A. and Hariharasubramanian, V. (1974) In Fraenkel-Conrat, H., (1969)—see general section.

Steitz, J. A. (1968) Identification of the A protein as a structural component of Bacteriophage R17. *J. mol. Biol.* **33**, 923–36.

Takebe, I. and Otsuki, Y. (1969) Infection of tobacco mesophyll protoplasts by tobacco mosaic virus. *Proc. Nat. Acad. Sci.* **64**, 843–8.

Thomas, P. E. and Fulton, R. W. (1968) Correlation of ectodesmata numbers with non-specific resistance to initial virus infection. *Virology* **34**, 459–69.

Van Vloten-Doting, L., Bol. J. F. and Jaspars, E. M. J. (1977) In Romberg, J. A. (ed.) (1977)—see general section.

Volovitch, M., Drugeon, G. and Yot, P. (1978) Studies on the single-stranded discontinuities of the cauliflower mosaic genome. *Nucleic Acid Res.* **5**, 2913–25.

Watts, J. W., Dawson, J. R. O. and King, J. M. (1980) *Viruses in tobacco protoplast: mechanism of infection.* J. Innes Inst. 71st Ann. Rep., p. 124.

Chapter 6

A'Brook, J. (1973) The effect of plant spacing on the number of aphids trapped over cocksfoot and kale crops. *Ann. appl. Biol.* **74**, 279–85.

Alphey, T. J. W. (1978) Chemical control of virus vector nematodes. In Scott and Bainbridge (1978)—see general section.

Barakat, A. and Stevens, W. A. (1981) Studies on the mode of action of inhibitors of local lesion production by plant viruses. *Microbios Letters* **16**, 7–13.

Bradley, R. H. E. (1963) Some ways in which a paraffin oil impedes aphid transmission of PVY. *Can. J. Microbiol.* **9**, 369–80.

Brandes, J. (1925) In Russell (1978)—see general section.

Brenchley, G. H. and Wilcox, H. J. (1979) *Potato Diseases.* HMSO, London.

Broadbent, L. (1964) Control of plant virus diseases. In Corbett and Sisler (1964)—see general section.

Broadbent, L. (1976) Epidemiology and control of tomato mosaic virus. *Ann. Rev. Phytopath.* **14**, 75–96.

Burgess, H. D. (ed.) (1981) *Microbial Control of Pests and Plant Diseases 1970–1980.* Academic Press, London and New York.

Carter, W. (1973) *Insects in Relation to Plant Disease.* 2nd edn., John Wiley and Sons, New York.

Cassells, A. C. and Long, R. D. (1980) The regeneration of virus free plants from cucumber mosaic and PVY infected tobacco explants cultured in the presence of virazole. *Zeitschrift für Naturforschung* **35**, 350–1.

Commoner, B. and Mercer, F. L. (1951) Inhibition of the biosynthesis of TMV by thiouracil. *Nature* **168**, 113–4.

Cooper, V. C. and Walkey, D. G. A. (1978) Thermal inactivation of cherry leaf roll virus in tissue cultures of *Nicotiana rustica* raised from seeds and meristem tips. *Ann. appl. Biol.* **88**, 273–8.

Duffus, J. E. (1971) Role of weeds in the incidence of virus diseases. *Ann. Rev. Phytopath.* **9**, 319–40.

Ebbels, D. L. (1979) A historical review of certification schemes for vegetatively propagated crops in England and Wales. *ADAS Quart. Rev.* **32**, 21–58.

Harrison, B. D. (1981) Plant virus ecology: ingredients, interactions and environmental influences. *Ann. appl. Biol.* **99**, 195–209.

Hollings, M. (1965) Disease control through virus-free stock. *Ann. Rev. Phytopath.* **3**, 367–96.

Hollings, M. and Stone, O. (1980) Production and use of virus-free stocks of ornamental and bulb crops: some phytosanitary and epidemiological aspects. In Ebbels and King (1980)—see general section.

Kaiser, W. J. (1980) Use of thermotherapy to free potato tubers of alfalfa mosaic, potato leaf roll and tomato black ring viruses. *Phytopath.* **70**, 1119–22.

Kanagaratum, P., Burges, H. D. and Hall, R. A. (1979) Integration of *Verticillium lacanii* and *Eucarsia formosa* for whitefly control. *Ann. Rep. Glasshouse Crop Res. Inst.*

Kassanis, B. (1949) Potato tubers freed from leaf-roll virus by heat. *Nature* **164**, 881.

Kassanis, B., Gianinazzi, S. and White, R. F. (1974) A possible explanation of the resistance of virus-infected tobacco plants to second infection. *J. Gen. Virol.* **23**, 11–6.

Kring, J. B. (1954) Cited in Smith and Webb (1969), in Maramorosch (1969)—see general section.

Lambert, I. F. (1981) Combating nematode vectors of plant viruses. *Plant Disease* **65**, 113–7.

Lawson, R. H. (1981) Controlling virus diseases in major international flower and bulb crops. *Plant Disease* **65**, 780–6.

Lobenstein, G., Rabina, S. and Van Praagh, T. (1966) Induced interference phenomena in virus infections. In Beemster and Dijkstra (1966)—see general section.

Moericke, Von V. (1954) Cited in Smith and Webb (1969), in Maramorosch (1969)—see general section.

Nene, Y. L. (1973) Viral disease of some warm weather pulse crops in India. *Plant Disease Reporter* **57**, 463–7.

North, C. (1979) *Plant Breeding and Genetics in Horticulture.* MacMillan Press.

Rast, A. T. B. (1972) M11–16, an artificial symptomless mutant of tobacco mosaic virus for seedling inoculation of tomato crops. *Neth. J. Plant Path.* **78**, 110–2.

Russell, G. (1978) *Plant Breeding for Pest and Disease Resistance.* Butterworth, London.

Shepherd, J. F. (1977) Regeneration of plants from protoplasts of PVX infected tobacco leaves. II. Influence of Viruzolen on the frequency of infection. *Virology* **78**, 261–6.

Simpkins, I., Walkey, D. G. A. and Neely, H. A. (1981) Chemical suppression of virus in cultured plant tissues. *Ann. appl. Biol.* **99**, 161–9.

Slack, S. A. and Shepherd, R. J. (1975) Serological detection of seed-borne barley stripe mosaic virus by a simplified radial-diffusion technique. *Phytopath.* **65**, 948–55.

Smith, F. F. and Webb, R. E. (1969) Repelling aphids by reflective surfaces. A new approach to the control of insect transmitted viruses. In Maramorosch (1969)—see general section.

Stirpe, F., Williams, D. G., Onyon, L. J., Legg, R. F. and Stevens, W. A. (1981) Dianthins, ribosome damaging proteins with antiviral properties from *Dianthus caryophyllus* L. (carnations). *Biochem. J.* **195**, 399–405.

Thresh, J. M. (1976) Gradients of plant virus diseases. *Ann. appl. Biol.* **82**, 381–406.

Thresh, J. M. (1978) The epidemiology of plant virus diseases. In Scott and Bainbridge (1978)—see general section.

Tomlinson, J. A., Faithfull, E. M. and Ward, C. M. (1976) Chemical suppression of the symptoms of two virus diseases. *Ann. appl. Biol.* **84**, 31–43.

Walkley, D. G. A. (1980) Production of virus-free plants by tissue culture. In Ingram and Helgeson (1980)—see general section.

Woodford, J. A. T., Shaw, M. W., McKinlay, R. G. and Foster, G. N. (1977) The potato aphid spray warning scheme in Scotland, 1975–1977. Proc. of the 1977 British Crop Protection Conference,—*Pests and Diseases*, pp. 247–54.

Chapter 7
Ball, E. M. (1974) *Serological tests for the identification of plant viruses.* Amer. Phytopath. Soc.

Brakke, M. K. (1967) Density gradient centrifugation. In Maramorosch and Koprowski (1967)—see general section.

Francki, R. I. B. (1980) Limited value of the thermal inactivation point, longevity in vitro, dilution end-point as criteria for the characterization, identification and classification of plant viruses. *Intervirology* **13**, 91–8.

Hamilton, R. I., Edwardson, J. R., Francki, R. I. B., Hsu, H. T., Hull, R., Koenig, R. and Milne, R. G. (1981) Guidelines for the identification and characterisation of plant viruses. *J. Gen. Virol.* **54**, 223–41.

Kado, C. I. and Agrawal, H. O. (1972) *Principles and Techniques in Plant Virology.* Van Nostrand, Reinhold and Co.

Koenig, R. (1981) Indirect ELISA methods for the broad specificity detection of plant viruses. *J. Gen. Virol.* **55**, 53–62.

Markham, R. (1967) The ultracentrifuge. In Maramorosch and Koprowski (1967)—see general section.

Milne, R. G. and Luisoni, E. (1975) Rapid high-resolution immune electron microscopy of plant viruses. *Virology* **68**, 270–4.

Murant, A. F. and Taylor, M. (1978) Estimates of molecular weights of nepovirus RNA Species by polyacrylamide gel electrophoresis under denaturing conditions. *J. Gen. Virol.* **41**, 53–61.

Murant, A. F., Taylor, M., Duncan, G. H. and Raschke, J. H. (1981) Improved estimates of molecular weights of plant virus RNA by agarose gel electrophoresis and electron microscopy after denaturation by Glyoxal. *J. Gen. Virol.* **53**, 321–32.

Roberts, I. M. and Harrison, B. D. (1979) Detection of potato leafroll and potato mop top viruses by immunosorbent electron microscopy. *Ann. appl. Biol.* **93**. 289–97.

Thomas, B. (1980) The detection by serological methods of viruses infecting rose. *Ann. appl. Biol.* **94**, 91–101.

Torrance, L. and Jones, R. A. C. (1981) Recent developments in serological methods suited for routine testing for plant viruses. *Plant Path.* **30**, 1–24.

Warmke, H. E. and Christie, R. G. (1967) The use of dilute osmium tetroxide for preservation of three dimensional crystals of TMV. *Virology* **32**, 534–7.

Index

177

180

Printed in the United States
By Bookmasters